BEFORE THE INDIANS

BJÖRN KURTÉN

Before
the Indians

ILLUSTRATED BY

MARGARET LAMBERT NEWMAN
AND HUBERT PEPPER

COLUMBIA UNIVERSITY PRESS   New York

Columbia University Press
New York   Guildford, Surrey
Copyright © 1988 Columbia University Press
All rights reserved

Library of Congress Cataloging-in-Publication Data

Kurtén, Bjorn.
  Before the Indians.

  Bibliography: p.
  Includes index.
  1. Mammals, Fossil. 2. Paleontology—Pleistocene.
3. Paleontology—North America. I. Title.
QE881.K793 1988   569'.097   87-18413
ISBN 0-231-06582-5   PA ISBN 0-231-06583-5

Printed in the United States of America

Hardbound editions of Columbia University Press are Smyth-sewn
and printed on permanent and durable acid-free paper

Contents

Introduction

WE DO NOT KNOW for sure when the first men appeared in America. What we do know is that vigorous people called Paleoindians flourished here at the end of the Ice Age, in the last millennia before the great transition of 10,000 years ago when the great ice sheets that had covered the northern part of the continent were finally vanishing. The Paleoindians are regarded as the ancestors of today's Indians.

What did these early men encounter in their new homeland? That is the question we hope to answer in the present book. As will be seen, they found a virgin continent, populated by a wondrously rich animal world—in some respects surpassing even the magnificent fauna of today's African savannas. They also found a tremendous variety of environments and scenery, ranging from the bitterly cold areas in the north to the subtropical worlds of Florida and the Gulf Coast, where the eruption of cold was hardly felt at all.

The great fauna of the Ice Age contained many elements of North American origin, but also animals that had immigrated, from time to time, from South America, from Asia and Europe, and even—ultimately—from Africa. (America has a long heritage as the melting pot of the world.) Here we trace the history of this fauna as it unfolded during more than three million years before the entrance of man in the New World. In geological parlance, this is equivalent to the final part of the Tertiary period and all of the Pleistocene, or Ice Age, epoch.

How can it be possible to reconstruct the past in this way? We wish to know, for instance, about the plants and animals of a certain region in the past, about its climate and physical environment, and about the time scale of its history. Many different branches of science converge in such work, and the first chapter describes how it is done. (There will be more about that in the running text of the enclosed portfolio of paintings.) Chapter 2 describes the scene before the onset of the glacial age, and the next chapter tells of the coming of the cold and of the earlier part of the Ice Age. Chapters 4–8 describe the panorama of the late part of the Ice Age as it was at the entrance of man.

In those latter chapters, various regions are described: the Ice Age looked very different in Alaska and Florida, in California or in the Great Plains, with quite distinct animals and plants. Many animals, however, were adaptable enough to survive in highly different environments—for instance, the scimitar cat *Homotherium* ranged from Alaska to Florida. Such species are presented in the context that appears most natural, for instance where they are especially common or where their presence is otherwise significant—like, the first appearance in Alaska of immigrants from Siberia.

The final chapter recounts the entrance of man and the final demise of the Ice Age fauna. Although close to us in the geological time-scale (10,000–15,000 years ago), the story is full of riddles, problems, and unanswered questions, posing new challenges to the student. The same, of course, is true for the earlier part as well. The proper response is imaginative and dynamic research. "Paleontologists," an anonymous writer once observed, "are ever ready to argue round a question that has more than one possible answer: long may this continue." Science is never "finished"—new problems arise, demanding new appraisals and new looks at what may long have been taken for granted. So it is with the study of the Ice Age. It is a healthy science.

This book grew out of another book. That book is *Pleistocene Mammals of North America*, written by Björn Kurtén and Elaine Anderson, and illustrated in part by Margaret Lambert Newman. At an early stage of that work, it was realized that a more popular account of the same material was desirable, and so it went into the planning stage soon after work on the earlier volume was started in 1970. *Pleistocene Mammals of North America* was the most important source text for the present book. Other sources and suggested reading are collected at the end of the book.

We would like to thank the numerous colleagues who helped us by giving information and access to material in their care. Special gratitude is due to Dr. R. Dale Guthrie for his assistance in developing a life restoration of Ice Age bison.

I

From Palo Duro
to Moonshiner Cave

BEGINNINGS

WE ARE STANDING at the rim of Palo Duro Canyon, Texas, its wide gash burning like a red wound on the drab landscape. This, to me, is where the story begins, to end more than three million years later and a thousand miles away in Moonshiner Cave, a punctured lava blister, hidden and inconspicuous beneath the sagebrush plain of Idaho. In between is the story of the Ice Age in America.

The Palo Duro is one of the places where we can recreate the American scene before the coming of the cold. At the same time, as we shall see, it shows the Ice Age to be just an episode within the vaster range of geologic time.

At Moonshiner, we are very close to our own time: the record starts at the end of the Ice Age, at a time when the first Indians were already on the scene.

The story in between, then, is that of the Ice Age—its convulsions and slow changes, its cold and heat, and, above all, its living beings, facing us out of the past in mute array. In myriad shapes, from diminutive to majestic, they move on the stage, live their lives, and sink into the earth that bore them. In our imagination they loom up as if called back to life: sometimes menacing, sometimes comical, often beautiful, and always intriguing: mysteries to be solved, riddles to be answered. Their story is one of great significance, but without moral or intent. Only at the very end does the cold light of human purpose fall across the stage, and swift, deadly shapes dash through the continent as if bent on self-destruction. At that point, in retrospect, it might seem that the entire story built up to a climactic tragedy.

But from where we stand now, more than three million years ago, such things might be thought of as in the distant future. With a thousand fingers the Palo Duro and its tributaries are clawing into the plain, cutting new slices from it, and each slice reveals a section of the past. Very innocent-looking, the little stream that is Palo Duro Creek may be glimpsed, as if protesting that it could not have had anything to do with carving this monstrous slit into the land of Texas. Don't be taken in. It has already dug itself 800 feet into the layered cake of strata that form the high plain of the Panhandle, uncovering beds of clayey rock that had rested in peace for more than 250,000,000 years.

Such incredibly ancient rocks form the base of the canyon's walls. They are a deep brick red with interbedded white bands of gypsum. These are the Permian Red Beds of Texas, far older than the oldest dinosaur. The oldest known reptile egg came out of such strata. At the time of their deposition this land was flooded by the sea. As the sea became landlocked, evaporation caused gypsum and rock salts to be deposited. This happened many times in untold millions of years.

Upon the Red Beds rest the varicolored rocks of the Triassic period. Mostly shales and sandstones, they glow with vermilions, pinks, and other tints. The greater part of the canyon wall is Triassic in age and, moving upward, we are already approaching the rim where we can see a great change occur. The Triassic rocks are left beneath and we are in a world of buff and dun-colored river sands and gravels. This is what the geologists term the Ogallala Formation, and it was deposited in Pliocene times, some 5 million years ago. A great stretch of earth history is missing here: the Triassic rocks are separated from the overlying Pliocene ones by an unconformity, or erosion surface. To sample that part of geologic history we would have to go elsewhere.

What we are looking for is the youngest part of the Pliocene, the record of the Blancan age: the last millions of years before the coming of the cold. We find these strata at the very top. In the grand geological perspective given by the sequence in the canyon walls, this final part seems little more than an insignificant episode. But that is deceptive. When you start looking at the "episode," it becomes an epic in itself.

To the southwest of the main canyon extend two tributaries, Little Sunday Canyon and North Cita Canyon. In this area, erosion has uncovered deposits of the Blancan age, and they are rich in fossil bones. The river sands of the lower Pliocene are topped by a hard, massive stratum of a rock called caliche—a deposit hardened by lime. Upon that rest sediments that were formed in a shallow basin some 2 to 3 million years ago, in Blancan times.

There was no canyon then, just the plain. The basin measured about three miles east and west, and one and a half miles north and south. In its northern part was a small permanent lake. Further south, temporary sheets of water accumulated in the rainy season, and evaporated in the dry. Such temporary lakes are called playas, and the Cita Canyon playa and lake sampled a very good record of the animals that lived in the area in Blancan times. The streams that accumulated the deposits and embedded the animal bones came in from the south.

The bones, it may be surmised, are of animals seeking the water, and dying in the area, perhaps of thirst, hunger, illness, or carnivore attack. The skeletons, picked clean and torn apart by coyotes and vultures, were then caught up pell-mell by the streams as the next rainy season came along, and deposited in the lake or playa beds. Centuries passed, and the remains of thousands and thousands of animals were sealed into the rocks as safely as into Davy Jones' locker.

Much later the drainage pattern was rejuvenated and the Red River and its tributary, Palo Duro Creek, started to cut themselves into the prairie of the Panhandle. Suddenly the locker was safe no more. Erosion was back at work, filing away at the Cita Canyon deposit, shamelessly exposing the skulls, teeth, and bones of the dead.

Now Little Sunday and North Cita Canyons are dissecting the Blancan playa, and at this point the fossil-bearing site was discovered by Floyd V. Studer and C. Stuart Johnston in 1935. The first great study of the fossil fauna was carried out by Johnston, but his untimely death only four years later cut it short. The study remains a mere torso, and although other writers have published material about the animals from Cita Canyon, no complete account is yet at hand.

We now know that the Blancan Age lasted millions of years. It may have begun as long as 4 million years ago and ended about 1.9 million years before the present (B.P. for short). The Cita Canyon fauna dates from the later half of this sequence. Other sites in many parts of the continent fill out the picture of the Blancan Age. Its history will be set down in more detail in the next chapter.

ICE

With the passing of the Blancan, we are entering the Ice Age, the Pleistocene Epoch in geological time. The last 1.5 million years of earth history have been dominated by a succession of great glaciations, during which almost all of Canada and the northern United States were covered by an immense land ice like that of Greenland and Antarctica today. A similar history was enacted in the Old World, and there is good evidence that glaciations were synchronous world wide.

The great northern icefields were matched in the south by the Antarctic inland ice. Also, smaller icefields developed in southern South America, and on the Australian island of Tasmania.

The idea of an Ice Age was first formulated by a Swiss scientist, Louis Agassiz, in 1837. His studies of ice-polished rocks and ice-laid deposits far outside the limits of present-day glaciers in his Alpine homeland led him to propound the concept of an age in which great ice sheets existed in what are now temperate areas. Observations in the British Isles confirmed his ideas. Later on he migrated to America (he was the founder of the prestigious Museum of Comparative Zoology at Harvard) and was able to identify traces of the Ice Age in this continent as well.

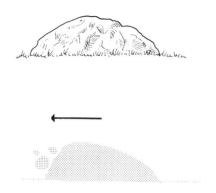

A rocky knoll is polished by the ice, forming a *roche moutonnée*. Arrow shows direction of ice movement. On the side facing the advance, striae caused by rocks frozen into the foot of the glacier show the precise direction of ice movement. The leeward side is steep and unpolished.

The Pleistocene Ice Age is by no means unique. Other ice ages have occurred at much earlier times in earth history. In fact, the penultimate ice age ended about 250 million years ago—or about the time when the lowermost rocks in Palo Duro Canyon were deposited. There was a still earlier one upwards of 600 million years ago. So it seems that ice ages tend to recur every 300 million years or so in earth history.

The Ice Age may be seen as the culmination of a very long history of gradually increasing cold, which had been going on for many millions of years. Warmth-loving plants and animals, once found far north—even in the Arctic—slowly withdrew to the south, and new types of organisms, equipped to withstand a harsher climate, evolved to take over the northern regions. Eventually, the ice sheets developed and drove almost all living beings out of the glaciated areas.

The icefields that covered the north of the American continent consisted of two major parts. One of these, the Cordilleran ice, formed on the mountains of the northwest. Further to the east, another ice sheet formed around a nucleus in the Hudson Bay area; this was the Laurentide ice. The latter icefield, by far the larger of the two, extended far enough to the west to coalesce with the Cordilleran ice, when glaciation was at its maximum, and so a single, immense barrier of ice was formed, extending from the Pacific right across to the Atlantic. Covering some 6 million square miles, it was even greater than the Antarctic ice of the present day, and may have been the mightiest ice sheet ever to exist on the earth.

As mementos of the land ice, tills and related types of glacial drift deposits blanket enormous areas of land that were once under the ice. The typical character of the tills is their unsortedness—they are clayey, sandy, gravely, with pebbles and boulders, in varying mixtures. They consist of material that was frozen into the ice, dragged along by it, and finally melted out of it.

At the margin of the ice-field such material accumulates in the form of a wall or

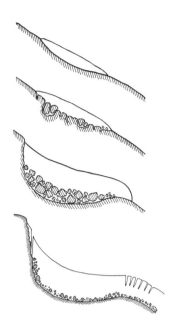

Birth of a glacier. As the mass of ice grows, it gradually attains a weight which causes the bottom layers to become plastic and start to "flow." Transverse cracks form in the brittle upper layers as the ice moves over a convex surface (bottom). Its head now forms a so-called corrie (or cirque) from which a glacier flows down.

terminal moraine, which may be a marked feature in the landscape. It denotes a shorter or longer pause of the ice margin.

Where a tongue of ice presses through a valley it tends to scour its sides into a characteristic U-shape, which we can now recognize in many formerly glaciated areas. In such ways the ice leaves its impress on the landscape, enduring long after the passing of the glacier itself. Even the sheer weight of the land ice, which could become up to two miles thick, is sensed by the earth: the underlying rocks are forced down by the enormous weight, and rise again, rebounding, when the ice is gone.

The surface of the inland ice is almost devoid of life. Very few animals will trek across it and those that do run the risk of falling into the crevasses that open and close with the movement of the ice. But the area outside the ice is habitable to man and beasts, even though it may be a bleak tundra. Although the surface may be ice-free in summer, great tracts are yet in the grip of the cold: the ice is hidden under our feet, the very ground is frozen throughout the year, except for its uppermost layer.

The presence of such permafrost in the past can now be recognized from various

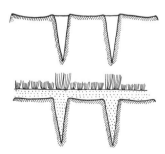

Ice-Wedges form as freezing water turns into ice-needles, forcing the sides of a crack apart. Long afterward, when the wedges are filled with sediment, the cracks may still be traced because the vegetation above them is especially luxuriant.

geological features, of which perhaps the so-called ice wedges are the most conspicuous. The freezing ground contracts, opening into ice-filled ditches forming quaint patterns on the surface. Later on, when the ice has melted, the cracks may become filled with other deposits, and now can be traced as fossil ice wedges. Sometimes, paradoxically, the one-time ice wedges stand out as lines of particularly luxuriant vegetation because the fertile soil that fills the cracks is deeper than that of the surrounding surface.

The relationship of the sediment called loess to glacial features has long been recognized. It consists of windblown dust, originally produced by frost erosion, then trapped by the sparse vegetation of the cold steppes; it may form deposits of great thickness. These are found in wide areas south of the erstwhile ice margin, and spread southward along the Mississippi and its tributaries. Some of the most fertile soils of the continent are of such origin.

So much ice was dumped on land—not only in North America, but elsewhere too. Europe and Asia were part covered by inland ice, and the icefields of Greenland and Antarctica were in place then as now. The southern extremes of South America, Africa, and Australia also supported glaciers of varying size. Where did all the ice come from?

Ultimately it came from the ocean: evaporation carried the water into the atmosphere, from where it was precipitated in the form of snow; and the snow, accumulating from year to year, was compacted into glacier ice. With the loss of so much water, the level of the ocean dropped world-wide by 300 feet or more.

As a result, many areas now covered by shallow seas were then dry land. To us the most interesting is Beringia, a wide, shallow platform connecting Alaska with Siberia. When the glaciation was at its maximum, Beringia formed a broad, mostly ice-free land bridge uniting the New World with the Old. So animals and plants, and man himself, could migrate from continent to continent, provided they were hardy enough to withstand the cold. And so the living world of North America was enriched, time and again, by immigration from Asia. (A land bridge was in existence in the area, intermittently at least, in pre-Ice Age times as well: at the beginning of the Blancan, for instance, brisk intermigration was taking place.)

Another path for such migrations was formed by the Central American isthmus, which has been in existence since Pliocene times. Of course, the movements went both ways: animals and plants of North America origin pressed into the sister continent of Asia and South America, embarking upon what was in many cases to become a spectacular scenario of relocation.

THE PENDULUM OF CLIMATE

Agassiz, and his nineteenth-century followers, at first conceived of the Ice Age as a single cold period. Toward the turn of the century, however, it became increasingly clear that the story was much more complicated.

In many places, a road cut may expose a succession of two till deposits, one on top of the other. The upper till will seem quite fresh, and some of its boulders will still show the striae or grooves produced when they were dragged by the moving ice across underlying rocks. In contrast, the lower till shows evidence of heavy weathering. Its boulders may have decomposed into soft lumps, and the other ingredients, too, have changed and rotted.

Such an ancient till, for which the name gumbotil has been coined, is generally separated from the upper till by an intervening layer which may show clear evidence of having been deposited in a warm climate. There may, for instance, be a layer of peat intercalated, containing the remains of plants that could only have grown in a warm or temperate climate. Such discoveries led to the conclusion that there have been several successive glaciations, with interglacials in between, when the climate was much the same as now. And remains of terminal moraines, in very different states of weathering and obliteration, helped to show the limits reached by the various glaciations.

To see the past as something other than just a jumble of facts and events, it must be given an organized form, a *gestalt:* it must be classified into periods, epochs, and ages, each of which is given a name and a profile of its own. Mapping the terminal moraines in eastern North America, from Montana to the Atlantic, Thomas C. Chamberlin, that great organizer and name-giver in American Pleistocene geology, called them historical ramparts which give us definite datum lines in the glacial epoch. On such moraines and tills, and on intervening "warm" deposits, four great glaciations and three interglacials have been recognized.

The glaciations have been named for the states in which particularly good evidence for their presence was found, and are called, from the oldest to the youngest, the Nebraskan, Kansan, Illinoian, and Wisconsinan. Names for the interglacials were taken from local sites: the Aftonian (between Nebraskan and Kansan) from peat exposed near Afton Junction, Iowa; the Yarmouthian (between Kansan and Illinoian) from spoils of a well at Yarmouth, Iowa; and the Sangamonian (between Illinoian and Wisconsinan) after Sangamon County, Illinois. Most of these names were given by Chamberlin and by another great pioneer, Frank Leverett. Chamberlin himself went on to even grander problems, finally to formulate far-reaching theories of the origin of the earth.

In this way, a first step was taken toward a chronology of the Ice Age. But it was still a chronology without any dates. The lengths of the various phases were unknown. Estimates could be made, of course, of the rates of sedimentation and weathering; and they suggested that the Ice Age had lasted about a million years. Although this was little more than an informed guess, we know now that the order of magnitude was right. The beginning of the Nebraskan Glaciation is currently set at about 1.5 million years.

DATING

The discovery of radioactivity gave us a new tool for measuring geological time. Radioactive decay goes on at a fixed rate, distinctive for each radioactive element, regardless of whether you freeze them or cook them, put them under pressure or into a vacuum. For instance, radioactive potassium, or K^{40}, disintegrates very slowly; it takes 1.3 billion years for one-half of a given amount of K^{40} to change into other elements. This is called the half-life of K^{40}. After 2.6 billion years, only one-fourth will remain unaltered; after 3.9 billion years one-eighth; and so on.

Potassium is abundantly present in many kinds of rocks and minerals, but only about $1/10000$th of the natural potassium is radioactive K^{40}. Its scarcity, and the fact that it decays very slowly, means that it does not cause dangerous radiation. The bulk of the potassium consists of the isotopes K^{39} and K^{41}, both of which are stable.

The decay of a K^{40} atom may lead to two quite different results. It may change into a calcium atom of the same atomic weight, and this happens to 89 atoms out of a hundred. But this is no good to the geochronologist, as he has no means of distinguishing the calcium thus produced from calcium that may have been originally present in the rock.

The remaining 11 percent of the atoms, however, turn into atoms of the noble gas argon, Ar^{40}. Now, argon (together with its more publicized but rarer sister gases, neon and helium) is also found in the atmosphere; but there it is a mixture of three isotopes, Ar^{40}, Ar^{38} and Ar^{36}, in the ratios 296/0.19/1. That makes it possible for the geochronologist to distinguish the radiogenic argon from such atmospheric argon as might have insinuated itself into his sample. Overcoming various sorts of snags, he finally arrives at a potassium-argon ratio, and is able to put a date on the origin of the rock under study.

The potassium-argon (or K-Ar) method is perhaps the most important of the "absolute" dating routines now in use, although it has turned out that only certain types of potassium-bearing minerals are suitable for dating. Its dating range is impressive, too, extending from about 200,000 B.P. back to and beyond the origin of the earth (some 5 billion B.P.). Moon rocks have been dated in the same way, and the results suggest that the age of the moon is of the same order of magnitude as that of the earth.

When we approach the end of the Pleistocene, however, such long-term methods become powerless, and we have to find others. The short-term method par excellence is radiocarbon dating. Like the potassium-argon method, it was developed in the years after World War II; but, apart from being based on radioactivity, its principle is very different.

Planets and stars move in an unending stream of visible and invisible radiation. Some of it is filtered away by the atmosphere; some, including visible light, reaches us at the bottom of the ocean of air. More penetrating than any others are the cosmic rays, whose energy is such that they knock pieces out of the atoms which they hit. As early as 1946, Willard F. Libby saw that the cosmic ray bombardment must produce radiocarbon in the atmosphere, and his prediction was quickly confirmed.

Radiocarbon, or C^{14}, is formed when an ordinary atmospheric nitrogen atom is hit by a radiation-induced neutron; it is then transformed into carbon with the same atomic weight (14). Radiocarbon, like ordinary carbon, is utilized by living plants and goes into their tissues; and since all animals live off plants, directly or indirectly, they will also contain radiocarbon.

As its name indicates, radiocarbon is radioactive, and it decays to form ordinary nitrogen again. This means that the production of new radiocarbon, going on all the time in the upper atmosphere, is balanced by the decay of old; and so the concentration of radiocarbon in the air, and in living organisms, remains unchanged. But as soon as an organism dies, its assimilation of radiocarbon comes to an end, and the amount of such live carbon in its body begins to dwindle.

This process, though slow when measured against our everyday concept of time, is much faster than the decay of, for instance, radioactive potassium: the half-life of radiocarbon is a mere 5,730 years. The time that has elapsed since the death of an organism is measured from the radioactivity of its carbon, and this is a good chronological tool for the last 50,000 years or so of geological time; in older samples the radioactivity is too low for efficient measurement. As we shall see, the entrance of man in North America, and his dramatic early history here, is now seen enacted on a scale of radiocarbon years.

Obviously, there is a gap between the maximum age datable by radiocarbon, and the minimum age datable by the K-Ar method. This is being filled by various other methods. One of the most efficient, and already in wide use, is the study of tracks made by disintegrating radioactive atoms in mineral crystals—the fission-track method. Another, the amino-acid racemization method, focuses on the postmortal changes of the optical properties of the amino acids in once-living tissues.

Geochronology is also served by study of the magnetic properties of rocks. When a rock is formed, its magnetic components are aligned by the earth's magnetic field. But the field reverses upon itself at irregular intervals in geologic time. Should this happen now, all our compasses would reverse their directions, magnetic north becoming south and vice versa. Explanations of this eccentric behavior are still very tentative.

Be that as it may, the result is that there are "normal" and "reversed" magnetic epochs (now called *chrons*) in the history of the earth, and the rocks formed in the different chrons can be recognized from their magnetic properties. The chrons have been dated, mainly by the K-Ar method. The current Brunhes normal chron (named for a famed geophysicist, like the other chrons) commenced 700,000 years ago, and was preceded by the reversed Matuyama chron, starting about 2.4 million years ago. Furthermore, chrons may contain much shorter "events," now styled *subchrons*, with temporary reversals of the field. Within the long Matuyama chron, for instance, two major subchrons of this kind have been found. The spacing of these *subchrons* gives each chron a profile of its own, and so helps to identify it.

There are, then, two independent stratigraphies of the alternation, or black-and-white type: one an alternation of cold and warm climates, the other an alternation of normal and reversed magnetic polarities. The two interfinger and balance out each other.

GLACIALS AND INTERGLACIALS: WHY?

Many theories have been proposed to account for the remarkable swing between cold and warm revealed by Pleistocene geology. They range from changes in the radiation of the sun, or the entire solar system passing through intermittent clouds of space dust, to changes in the positions of the poles and continents of the earth; and so far no consensus has been reached. But there are some important recent advances which may point to a solution.

It is rare to find on land long sequences of deposits which record, in undisturbed stratigraphic superposition, long series of climatic events. But there are many places at the bottom of the sea in which sedimentation has gone on, virtually undisturbed, for many millions of years. Day after day and year after year, myriad shells of microscopic dead animals are sinking down, settling at the bottom, and becoming entombed in its sediment. These microfossils can be identified and their climatic significance evaluated. Thus they record the climatic conditions that prevailed in their lifetime.

New deep-coring techniques have been developed to extract long cores of such sediments. The abundant microfossils tell us of the climatic fluctuations, and they can be dated by the magnetic polarity history which is also read in the cores.

Such studies show that the classical fourfold glaciation picture of Chamberlin and Leverett (Nebraskan-Kansas-Illinoian-Wisconsinan), while valid in a broad sense, is an oversimplification. Within the last half million years, for instance, the climate curve shows at least five major cold phases and various minor ones, interspersed by relatively brief warm intervals: in other words, there is a major periodicity of about 100,000 years, superposed on which are short-term periodicities. A group of scientists associated with a project for the study of the earth's climatic history (CLIMAP),

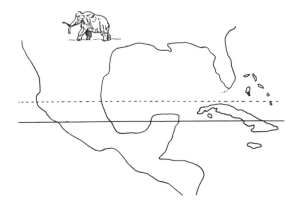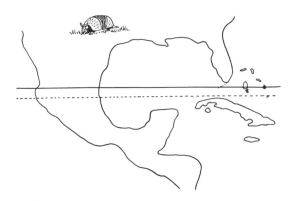

Shifts in the position of the Tropic of Cancer are correlated with climate. Left shows Tropic at extreme southern position during glaciation; right, Tropic far to the north, climate is warmer. Dashed line shows present-day position.

J. D. Hays, J. Imbrie, and N. J. Shackleton, showed that the curve may indeed be analyzed as a combination of wavelengths, somewhat like a phonogram recording the simultaneous playing of a number of pure notes. The three longest (and most powerful) wavelengths are 100,000, 40–42,000 and 19,500–24,000 years, respectively.

This interesting combination rings a bell (or, in fact, three) in the mind of the geochronologist. In the 1920s, a Serbian engineer and mathematician, M. Milankovitch, and two German climatologists, W. Köppen and A. Wegener, advanced a new theory of the ice age. It has been called the "astronomical theory," because it relates the history of the ice age to certain regular changes in the earth's motion around the sun. These are three in number.

In 1609, the German astronomer Johannes Kepler showed that the planetary orbits are elliptical. But the shape of the earth's orbit varies from elliptical to nearly circular and back to elliptical with a periodicity of about 100,000 years. And that is the longest wavelength in the Pleistocene curve.

Second, it is known that the inclination of the earth's axis against the plane of the orbit varies regularly. At present, for instance, it posits the Tropic of Cancer at 23°27' North (just north of Havana, Cuba). But at its northern extreme it is at 24°36' (in the Florida Keys); and at its southern extreme it is at 21°39' (crossing the point of the Yucatán Peninsula). The period of this change is 41,000 years: the same as the second longest wavelength for the Pleistocene.

Third, there is the precession of the equinox. At present the earth is closest to the sun during the midwinter of the northern hemisphere, but 10,500 years hence this will occur at midsummer and after another 10,500 years the situation will be back to the present. This period, then, is 21,000 years, which fits in very well with the third of the Pleistocene wavelengths. Hays, Imbrie, and Shackleton, indeed, consider the agreement significant; in fact, as early as the 1950s, Cesare Emiliani of Miami University showed that the climatic curves agreed with predictions based on the astronomical theory.

Though these changes do not affect the total amount of radiation received by the earth from the sun, they do affect its distribution over the earth's surface, in some instances producing cool summers and mild winters, in others hot summers and cold winters. Except for the 41,000-year periodicity, the effects are antithetical on the northern and southern hemispheres; but that on the northern half of the globe with its concentration of land masses, where continental ice sheets can form, is the most important.

Although the astronomical theory may turn out to be a good explanation for the alternation of cold and warm episodes, it cannot be the ultimate explanation of the Ice Age. The same changes in the earth's orbit occurred before the Ice Age without resulting in glaciation. Thus there must be some other cause which, so to speak, sets the stage; many theories have, again, been proposed but in this case we may be even further from a consensus. A phenomenon of major importance may be "continental drift," the movements of the continents relative to each other and to the poles of the Earth. These movements are much too slow to influence the glacial-interglacial cycle, but the long-term 300-million-year periodicity is a very different matter. During the Tertiary, the North Pole moved, ultimately from the Pacific, into the Arctic Sea basin. As a result, very large land masses were brought into the subarctic regions, and this may well have set the stage for an Ice Age.

One feature stands out very conspicuously in the climate curve. The really warm intervals, or true interglacials, appear as brief episodes in a massively cold-dominated scenario. In fact it has been possible to measure the lengths of some of the interglacials. In many lakes, the bottom sediments show a clear lamination into annual layers, with the seasons marked by the flowering and death of microscopic plants called diatoms. Two ancient lakebottoms in Germany, for instance, have yielded the lengths of the last interglacial (about 11,000 years) and the penultimate one (about 16,000 years). And the climatic history of the last 10,000 years in earth history shows that we even now are living in an interglacial which is rapidly nearing its end. So, unless man manages to cause some radical changes in the trends of climate, a new glaciation appears to be just around the corner.

FOSSILS

We have looked at some of the chronological tools used by the student of the Ice Age. But the first and original tool is biostratigraphy: the placing of fossils into their correct order in time, as revealed by their position in superposed deposits.

Cita Canyon, a tributary of the Palo Duro, has already shown us one way in which

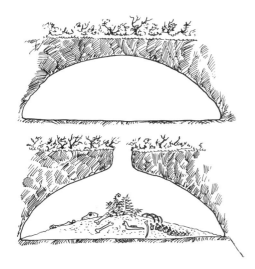

Diagram shows lava blister as formed, and punctured by rock-fall opening an entrance. Beneath the entrance forms talus cone containing bones of trapped animals. Luxuriant fern growth seems to reduce the distance to the bottom.

fossils may be preserved. The animal remains were deposited by running water. Moonshiner Cave in Idaho is also rich in fossils, but for a different reason. It is a small cave just beneath the surface of the lava plain, and it was formed as a great blister of gas in the lava flow when it was still bubbling hot. Its top has caved in, opening an entrance a few feet across. From that you can look down at the rubble on the floor, less than ten feet beneath. It really looks nearer because there is a luxuriant bed of ferns there, strikingly different from the bleak sagebrush vegetation of the plain. In winter, the cave looks even shallower as a cone of snow rises to within a few feet of the entrance. It will look like a perfect hiding-place to a smart animal. But not even the smartest animal can get out again, unless it can fly—or has the prescience to put down a ladder.

With a dead and decaying body in place, the trap is doubly baited. What an irresistible lure this has been to weasels, wolverines, foxes, and coyotes over the years is clear from the incredible number of bones found by Dr John A. White of Pocatello, who excavated the cave. Of the weasel only, more than 400 skulls were found.

He also found traces of man—to be sure, not in the form of bones. The cave, however, contained the remains of a still (hence the name "Moonshiner Cave").

The animals of Moonshiner Cave are not the same as those that lived millions of years ago in Texas. Life is dynamic, and they have changed—although some of those from the lava cave are descendants of some of the Cita Canyon species. A long, long time intervenes between the two faunas: the time of the Pleistocene. The creatures of Moonshiner Cave lived after the Pleistocene, in the Recent epoch of earth history, spanning the last 10,000 years or so, and the radiocarbon in their tissues is still active. That of the ancient Cita Canyon animals, which lived before the Pleistocene epoch, is long dead.

Its carbon dead or alive, few things probably look deader to a layman than a fossil bone. And for some scientific purposes it might well be just that. To the stratigrapher, trying to unravel the sequence of geologic deposits and their correlation from place to place, the fossil is simply a time marker. If he finds remains of the same species, say the miniature horse *Nannippus phlegon,* in deposits in Kansas and Florida, he concludes that the deposits are approximately of the same age (in this case, Blancan) or, in other words, can be correlated. For him to be able to do that, it does not really matter if it is a horse, or a snail, or a Clovis javelin point. Anything that is correlatable is grist for the stratigrapher's mill.

But to a paleontologist (the name is coined from three Greek words meaning "ancient," "life," and "study") the fossil is a glorious thing, full of intrinsic significance. It is a cipher letter from the past, crying to be decoded. His is the thrill of knowing that what lies before him is part of the stream of life, and it is up to him to find out its place within that vast unhurried flow.

It was in North America that paleontology was transformed from a sedate academic pursuit into a dynamic science charged with adventure and passion. Its figureheads in the late nineteenth century are the two great pioneers and rivals Othniel Charles Marsh and Edward Drinker Cope. Marsh, of Yale, cold-eyed and resentful; Cope, of Philadelphia, impetuous and outgoing; both great as scientists, and great as haters. Their spectacular fossil finds were front-page news in the nineteenth century, and so, unfortunately, were their rivalry and race for priority, resulting in innumerable scandals. The feud has passed into history, but their legacy is the first great collection of American vertebrate fossils: Cope and Marsh, and their associates, were the first to open up the marvelous fossil fields of the West. Of the two, only Cope came to do any serious work on Pleistocene paleontology; their main interest was in earlier parts of the earth's history. So it fell to later generations to chart the life of the ice age, and the work is still going on.

It was in North America, too, that life restoration of the distant past became an art. Student-artists like Charles W. Knight and R. Bruce Horsfall set wholly new standards for this type of work. In their pictures, which have been reproduced over and over again, the creatures of the fossil beds come to miraculous life. But their traditions are by no means forgotten, as anybody visiting the National Museum of Natural History in Washington, D.C., can verify from J. H. Matternes's long series of murals.

A restoration of a fossil mammal is based on meticulous anatomical studies. It begins with assembling the skeleton, assuring that we get the correct proportions of the various parts of the body. The skeleton will also tell us much about the typical stance of the animal. Few things are more characteristic than the poise of the head—proudly raised in the wapiti, broodingly thrust forward in the bison, hanging in the bear. A great deal of this can be read out of the shape of the skull and the joints of the neck. Even more, sometimes, is told by antlers or horns spreading out from the skull. They are organs of display, and an animal confronting you will use them to maximum effect as unerringly as if it had practiced before a mirror. You may have only the dead bones of the head, and yet you will be able to pose them the way the creature itself would have done.

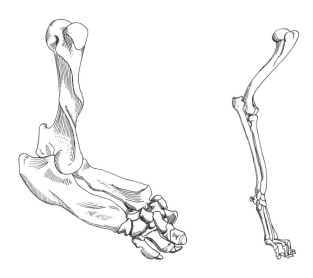

Heavy limb bones of slow-moving ground sloth contrast with slim bone structure of fleet-footed coyote.

Again, hands and feet reveal the secret of their poise long after death. The long slim hand-bones of a wolf tell as eloquently of a tiptoe racer, just as the square-built ones of a bear tell of a flat-footed walker. The back, straight or arched; the tail, rigid or lashing; it is all there in the bones.

Now flesh is added to bone. The various parts of the skeleton were linked together by muscle and tendon, and they, too, have left their mark on the bones. See the ponderous arm bone of the giant ground sloth, with all its crests and scars for the insertion of bulging muscles, and compare it with the slim and supple bone of the coyote, built for speed not power.

Finally, the hide and fur are added; and it is here, unfortunately, that we have to fall back on hypothesis. We do know something of the animal's habits and probable environment, and this may give some suggestions as to its color, but what are they worth? We know that steppe animals are often yellowish and forest animals usually darker, but there are no keys to the exquisite patterns of red pandas and ocelots. We may suspect that nature has always been far more inventive then the artist. After all, if zebras were extinct, who would have the imagination to dream up a pajama-striped horse?

LAND-MAMMAL AGES

Study of the immensely long and complicated geological history of the earth has been based on a very few simple principles. The first one was formulated by Nicolaus Steno (a Latinized version of his Danish name, Niels Stensen) in the seventeenth century. It states that, in a series of deposits, the lowermost are the oldest and the uppermost the youngest: this is the Law of Superposition. The second stems from a discovery, about a century later, by the French naturalist Jean Etienne Guettard. He was the first to realize that strata of a given age could be recognized on the basis of their contained fossils, and that such strata, or formations, could be traced over great geographic distances. Thirdly, the British student William "Strata" Smith, found that such rock units, each with its unique fossil content, were always stacked upon each other in the same order. At any site, from the lowermost (and thus oldest, as Steno had taught) to the uppermost and youngest, the arrangement was invariably the same, like the pages of a book. To carry the analogy further, it would be a very old book, with different pages missing in different copies: to get the complete story, we have to piece together innumerable book fragments. Thus was born the science of stratiography, "that branch of geology which treats of the formation, composition, sequence, and correlation of the stratified rocks as part of the earth's crust," to quote the definition of the American Geological Institute.

From such studies resulted a history of the earth in which all the geological strata were placed in their correct order of origin. The corresponding time intervals are classified as follows. The longest are called Eras (e.g., the Cenozoic Era). Eras are divided into Periods (e.g., the Cenozoic into the Tertiary and Quaternary Periods); Periods into Epochs (e.g., the Quaternary into the Pleistocene and Holocene Epochs); and Epochs into Ages (e.g., the Pleistocene into the Early, Middle, and Late Pleistocene Ages). Each of these divisions is characterized by its own set or sequence of fossil organisms. This is the standard geological time scale, to be found in any textbook.

Now the history of the earth was entirely based on marine deposits, and thus on the history of life in the sea. When continental deposits are intercalated between marine ones, they can be dated within this framework. But many continental deposits, especially from the Age of Mammals (the Cenozoic), have no stratigraphic connection with the strata laid down in the sea, and so their relationships to the marine time-scale remained doubtful. You might take, as an analogy, the division of Western history into Classic Times, the Middle Ages, and the Modern Epoch. This scheme works well in Europe but is of little use if we study, say, the history of China. Much better, then, to construct a local scheme for Chinese history, and work within that framework. Later on, we can try to find out how it correlates with the European one.

A summary of American faunal history during the last 3.5 million years. The time scale begins at bottom and ends at top. Symbols left indicate first appearance of some guide fossils: the Blancan age begins with the evolution of modern-type horses (genus *Equus*); the Irvingtonian age with the immigration of mammoths (genus *Mammuthus*); and the Rancholabrean age with the immigration of bison (genus *Bison*). The sabertooth cat, *Smilodon fatalis*, evolved at about the middle of the Irvingtonian age. A series of radiometrically dated Pearlette Ash falls intercalated into the sedimentary column help to clarify the chronology, and so do the swings in magnetic polarity as shown along the right border of the diagram—black denotes times with normal polarity, white times with reversed polarity. The timing of some important fossil-bearing sites is also shown, as well as the chronology of the four glaciations and three interglacials recognized in America.

From Palo Duro to Moonshiner Cave

Mil. yrs.	First appearance	Site	Glacial or Interglacial	magn eps.	
		Moonshiner Cave			BRUNHES EPOCH
			WISCONSINIAN GLACIATION		
		Reddick	Sangamonian Interglacial		
	Beginning of Rancholabrean Age		ILLINOIAN		
0,5			GLACIATION		
	Pearlette Ash "O"	Slaton Quarry	Yarmouthian Interglacial		
		Irvington	KANSAN		M A T U Y A M A E P O C H
		Cumberland Cave	GLACIATION		
1,0					
	Pearlette Ash "S"		Aftonian Interglacial		
			NEBRASKAN GLACIATION		
1,5					
		Port Kennedy Cave			
	Beginning of Irvingtonian Age	Curtis Ranch			
2,0	Pearlette Ash "B"				
		Mt. Blanco			G A U S S E P O C H
2,5		Cita Canyon			
		Sand Draw			
3,0		Rexroad			
		Hagerman			
3,5	Beginning of Blancan Age				GILBERT EPOCH
	End of Hemphillian Age				

In the same way, paleontologists decided to construct a separate, independent (and somewhat informal) framework for the history of land faunas: the provincial Land-Mammal Ages. As in the marine succession, these Ages are based on organisms (mostly mammals) and their succession through time. But they are not worldwide: each biogeographic area has its own sequence of Land-Mammal Ages. Neither are they grouped into Epochs or Periods. This informal and flexible approach has proved most useful. Just like the Ages of the standard geological time scale, each land-mammal age has a characteristic association of life forms which sets if apart from earlier and later ages. The Blancan age, for instance, may be defined as the time from the first appearance of modern-type horses to the first appearance of mammoths in North America.

The system was first introduced in North America, where it now covers the entire Age of Mammals. We only need to learn a few of the names—those of the four last ages. The first, which overlaps with the late Miocene and early Pliocene of the marine sequence, is the Hemphillian Age. The second, with which our story commences, is the Blancan Age (late Pliocene). Third comes the Irvingtonian Age (early and middle Pleistocene), fourth the Rancholabrean Age (late Pleistocene).

EVOLUTION

With the passing of time, the living world changes. The animals now found in the Texas Panhandle are not the same as those of Cita Canyon. Those that left their remains in Moonshiner Cave less than 10,000 years ago are different from those that lived in Idaho three million years earlier and whose bones are found at the Hagerman site. True, some mammals living today are so close to their Blancan ancestors that we refer to them as belonging to the same species. Yet the great majority are different, although they may resemble Blancan species so much that an ancestor-descendant relationship is indicated, and in many cases verified by discoveries of intermediate forms in Irvingtonian and Rancholabrean deposits.

If we look at shorter steps in time, for instance from the late Irvingtonian to the early Rancholabrean, the relationships are even closer. Many species pass through without any real change. Others have changed so much that the ancestral and the descendant form must be referred to as different species.

Looking at individual lines of descent, there is a great variation in the way changes occur. There are lineages showing a gradual, ongoing change, with ancestral and descendant species grading insensibly into one another (a good example is the history of the muskrat, discussed in chapter 3). In such cases, evolution has proceeded at a fairly steady rate for a long period of time. In other cases, there is a distinct gap between the putative ancestor and the descendant. There are two possible explanations for that. One is that the rate of evolution was stepped up markedly for a brief period, leading to the rapid rise of a new species. The other is that the new form evolved somewhere else, and then immigrated into the region we are studying, ousting and replacing the old form. If a species is widely distributed, local populations will diverge from each other, evolving into somewhat different subspecies ("races"). One of these might be cut off from the others by a geographic obstacle (for instance, an ice sheet), and continue to evolve into a distinct species. When the obstacle is removed (the ice sheet melts) it could spread back and take over. While this may be a good explanation in many instances, I would also agree that the stepped-up rate model may apply in certain cases.

It might be said, then, that there are three different modes of evolutionary rates, exemplified in the history of Ice Age mammals. Mode (1) is stasis over a long time—little or no change, as in those Blancan species that have survived to the present day. Mode (2) is gradual progressive change, going on for a long time and leading through two or more species-stages. Mode (3) is the episodic rapid change, too rapid to be normally shown in the fossil record. These models have been much discussed in recent arguments over evolutionary theory. In one current interpretation, the punctuated-equilibrium theory, mode (2) is considered insignificant, so that evolution would consist essentially of modes (1) and (3), stasis punctuated by episodes of rapid change. What is known of the history of ice-age mammals suggests that this is an oversimplification and that evolutionary rates are highly variable, ranging from no perceptible change to very rapid change. Such a conclusion was indeed reached in 1944 by George Gaylord Simpson, the first to study evolutionary rates in detail. His book *Tempo and Mode in Evolution* initiated a new era in the paleontological study of evolution.

Examples of the various evolutionary modes will be set forth in later chapters. However, it is only rarely that the information at hand is good enough to reveal all the pertinent details. Most known sequences are interfoliated by gaps. Still, as collecting and research goes on, the gaps tend to diminish and vanish. The record will never be complete, any more than we can count all the stars in our galaxy, but it certainly is good enough to prove the truth of evolution many times over.

In its modern version, evolutionary theory holds that evolution results from the interaction of mutation (or what Darwin called spontaneous individual variation) and natural selection. Mutation is random, as far as the needs of the organism are concerned (constraints do exist, for instance, on the molecular level). Selection is oriented: it may, for instance, promote the status quo or lead to a recombination of the genetic material. Thus, both chance and antichance are factors in evolution. This was the essence of Darwin's theory, presented to the world in 1859 (incidentally, he also knew that rates of evolution vary). But both mutation and selection are extremely complex and manifold processes, not to be expressed in simple catchwords. Even a rough overview, from the molecular to the organismal level, would become a book in itself. The suggestions for further reading at the end of this book will direct the reader to a number of comprehensive texts.

II

A World at Sunset

THE BLANCAN AGE

THE BLANCAN LAND MAMMAL AGE lasted from about 4 million to 1.9 million years B.P. It comprises the later part of the Pliocene Epoch in earth history, and the Pliocene was the last in the long sequence of epochs which make up the Tertiary Period. It was followed by the Pleistocene Epoch, the time of the Ice Age.

The beginning of the Blancan age is marked by the appearance of a horse of modern type (the American zebra, discussed in more detail below) and by temporary emergence of the Bering land bridge, with resulting intermigration between the New World and the Old. Similarly, it was brought to an end by another burst of migrations, ushering in the Irvingtonian faunas.

There are no traces of extensive inland ice from the Blancan of North America. But the climate apparently was cooler than in earlier Tertiary times, and there probably were continental ice sheets in other parts of the world. In Antarctica, inland ice formed many millions of years before the Blancan. And there is evidence of climatic oscillations during the Pliocene, analogous to glacials and interglacials but of lesser magnitude. There were times when big glaciers developed in mountainous country, and their memory remains in the form of till deposits. Some of these have been dated.

In Alaska, an early glaciation occurred about 3.6 million years B.P., or around the beginning of the Blancan, and there are traces of even older Alaskan glaciations back to some 8 million years, in the late Miocene. But these glaciations were local affairs.

The ice came later to the Rocky Mountains and the Sierra Nevadas. An old occurrence is the Deadman's Pass Till, bracketed between 3 million and 2.7 million years B.P. from underlying and overlying lava flows. A second cold phase, the Sierran Glaciation, is recorded by the McGee Till, which overlies a basalt flow dated at 2.6 million years B.P. But although these and other documents testify to episodic cooling, the impact remained local and restricted to mountainous areas. In the fauna and flora of the Blancan, there is little evidence of cooling, such as cold-adapted species, except perhaps toward the very end.

There are about thirty sites which have yielded good local faunas of Blancan age, and they are widely distributed in space and time. One of the most important is the Vallecito Fish Creek region, which is situated in the Anza-Borrego State Park in the Colorado Desert of southern California, 50 miles northeast of San Diego and 35 miles southwest of the Salton Sea. The sediments, derived from the bordering uplands and from the ancestral Colorado River, have since been tilted and dip at 24–25 degrees. The total area of fossil-bearing sediments exceeds one hundred square miles and their total thickness is over 12,000 feet. These strata cover all of the Blancan and much of the preceding (Hemphillian) and succeeding (Irvingtonian) ages. Paleomagnetic studies have established a detailed chronology for this pile of deposits, and as early as in 1968—fourteen years after the discovery—Drs. Theodore Downs and John White reported over ninety species found here. It is an incredible storehouse of successive faunas, with the material properly docketed, tier by tier. The Vallecito will become the standard of reference for Blancan history in southwestern North America.

Another famous site lies in southwestern Idaho, across the Snake River from the town of Hagerman, where a fabulous "horse graveyard" was discovered in the 1920s by Elmer Cook, a resident of Hagerman. Here, expeditions from the Smithsonian Institution in Washington, D.C., collected an immense material in the early 1930s, and later expeditions have augmented the harvest. At present, more than 300 localities in these beds have yielded more than fifty species of mammals from the early Blancan.

The San Pedro Valley in the southeastern corner of Arizona, again with a magnetically and radiometrically dated pile of sediments spanning all of the Blancan, is still another treasury of fossil mammals. In Texas are the Mt. Blanco and Cita Canyon faunas, both in the Panhandle area. That of Mt. Blanco (type locality of the Blancan) was discovered in 1889 by W. P. Cummins, whose collection was studied by Cope. At present, 45 mammal species are known from here.

In Nebraska, Broadwater in Morrill County and Sand Draw in Brown County are the two principal sites. Both probably date from the later part of the Blancan.

Thanks to the work of the late Claude W. Hibbard, Meade County in south-central Kansas is a classical area in the study of Blancan history. Repeated cycles of sedimentation and erosion have created an intricate stratigraphic patchwork which, though difficult to unravel, has been rewarding because of the rich harvest of fossil mammals at many sites. Most deposits are riverlaid but there are also filled-in artesian springs, where the animals came to drink and were entrapped in the quicksand. Dating is based on volcanic ash falls, the so-called Pearlette Ashes, which have been studied radiometrically, and on paleomagnetic work. The richest of the faunas is that from Rexroad, early Blancan, with 50 species of mammals.

In Florida, too, Blancan faunas are known, the outstanding one being that from Santa Fe River in Gilchrist County. Like many other Floridian find-spots, this is a river-bottom site. No absolute date is known but the fauna most closely resembles that of late Blancan sites elsewhere. There is also at least one late Blancan fossiliferous cave or fissure fill in the vicinity of Haile in Alchua County.

In addition to these localities, good Blancan faunas are known from Washington, Nevada, Colorado, and South Dakota. Coverage is very good west of the 95th meridian—coinciding approximately with the eastern state lines of the Dakotas, Nebraska, Kansas, Oklahoma and Texas—while the eastern part of the United States is a blank, except for Florida.

The great majority of the fossils come from water-laid open-air deposits, and the animal communities represent stream or pond, marsh and meadow, valley slope and upland. Mostly, the climate seems to have been more equable than at the present day. The varied animal life of Vallecito, for instance, points to conditions very different indeed from the harsh desert climate of today. At Hagerman, the flora and fauna—fish, frogs, water snakes, water and shore birds, and a variety of mammals—certainly suggest a warmer and more humid climate than that now prevailing in the area: the surroundings, now treeless, were forested, and there were streams, beaverponds, and lakes. At Sand Draw in eastern Nebraska, the fossils indicate a lower elevation than today; the area was watered by winding rivers with numerous lakes, and the climate was more equable and humid than now. So the overall picture of the Blancan is one of lingering Tertiary warmth, only sporadically interrupted by cooler phases in which glaciers were formed in the higher mountain ranges. Also, there was nothing like the climatic zonation of today.

A first impression of the animal life of the Blancan would probably resemble the one you get in a modern African savanna. As in Africa, there were great elephantlike animals. There were pigs and antelopes, and large herds of zebras. Lion-sized cats preyed upon the large herbivores, while the fleet-footed antelopes were pursued by cheetahs. But when we take a closer look at these creatures, the resemblance to present-day animals, in most cases, tends to dwindle. The elephants are not real elephants, but mastodons; the pigs turn out to be peccaries; the antelopes are pronghorns; and the big cats are sabertooths. Also, there are many animals which differ completely from anything seen in Africa.

In a few instances, though, the resemblance is very close indeed. The cheetah is a real cheetah. And those zebras are real zebras, too.

AMERICAN ZEBRAS

There are, or have been in recent times, "wild" horses in North and South America. But they are not wild in the technical sense of the word, for they are descended from domestic horses; they are "feral." Truly wild horses today occur only in the Old World; the zebras of sub-Saharan Africa; the asses of northern Africa; the hemiones or half-asses of Asia. There is also the near-extinct Przewalski horse of Asia, but it is uncertain whether that race survives in its original form, for it may have crossed with domestic horses. In any case, for nearly ten thousand years before the coming of the Europeans, no horse lived in the Americas.

In spite of this, North America was the center of horse evolution for most of the Tertiary. When speaking of horses in this sense, we mean all the members of the zoological family Equidae; and their American history goes back some 55 million years, to the late Paleocene.

We probably would not recognize the Paleocene "dawn horse" as a horse at all. It was a forest-living creature the size of a fox terrier, with four toes on the front feet and three on the hind, and its skeleton and teeth were highly different from those of a modern horse. Yet it is linked to its living descendants by an unbroken series of evolutionary stages, and this forms one of the best-known and rightly famous examples of evolutionary documentation.

The path from dawn horse to modern horse was by no means a "straight" one. At every stage, the animals were adapting to their current mode of life, not to one that might present itself a million years hence. The early descendants of the dawn horse were three-toed forest horses, with spreading feet suited to soft ground and low-crowned teeth suited to browsing. Later on, when the climate became dryer and grasslands took over in many areas, some of the forest horses gave rise to three-toed steppe horses with feet adapted to hard ground and higher-crowned teeth adapted to a grass diet. Finally, certain three-toed steppe horses gave rise to the modern one-toed steppe horses.

Each of these four stages, beginning with the dawn horse, is also represented in the Old World. But the evolutionary links between them have only been found in North America. So it is clear that they evolved here, and the Old World horses are migrants whose origin lay in North America.

The one-toed stage was reached just before the Blancan, and the appearance of the modern genus, *Equus,* is one of the time-markers which help us to fix the lower boundary of the Blancan age. However, one group of three-toed steppe horses, *Nannippus,* survived alongside with *Equus* in the Blancan. The name, signifying "small horse," is apt: these gazelle-horses reached the height of a Shetland pony at most, but were much lighter in build, with long, graceful legs.

The American zebra has been known to science since 1892, when Cope published his description of *Equus simplicidens* from Mt. Blanco. The name, "simple-toothed," alludes to the enamel pattern of the cheek teeth, which is simple in comparison with that of many other horses. Later on, other remains were unearthed, in California, Colorado, Florida, Kansas, Nebraska, Nevada, Texas, and Washington, but by far the greatest sample comes from the Hagerman horse quarry of Idaho. In 1930–34, the Smithsonian Institution expeditions found no less than one hundred and thirty skulls, fifteen skeletons, and many other bones of this animal; and the quarry has remained productive. John A. White states that this amazing site ranks among the four most outstanding fossil deposits in North America. (The others are Rancho La Brea of the late Pleistocene, to be described in a later chapter, and two older sites, Agate Fossil Beds National Monument in Nebraska and Dinosaur National Monument in Utah.) Unfortunately, a prodigious amount of bones collected by amateurs have been lost to science.

In 1944, Paul McGrew of Laramie, Wyoming, suggested that *Equus simplicidens* was in fact a zebra. This has since been confirmed and experts now agree that it is closely related to the living African species called Grévy's zebra *(Equus grevyi)*. It is the largest of today's zebras, standing up to six feet at the shoulders. It has large ears, somewhat reminiscent of a donkey, and a long, narrow muzzle. The black stripes are narrower and more densely spaced than in other zebras, which gives the animal a greyish look from a distance. A striking feature is a longitudinal black stripe along the back, separated from the vertical stripes by a white band. We do not know whether this pattern, or anything like it, was present in the American species, but the long narrow muzzle is the same. Grévy's zebra has a more northern distribution than other zebras, and is found in the northeastern part of a sub-Saharan Africa, up to southern Ethiopia and Somalia. It lives in small herds which are led by a stallion. The other living zebras are more distantly related to the *simplicidens-grevyi* group.

McGrew believes that zebrine horses migrated to the Old World at a very early stage, there to evolve into other types of horses—asses, hemiones, and true or "caballine" horses—of which some then remigrated across the Bering area into North America. This may well be true, although details of the story are not yet known. What we do know is that *Equus* appeared in the Old World at about the same time as in America, and soon spread across Asia, Europe, and Africa, gradually replacing the old three-toed horses which had migrated there in the Miocene. We also know that true asses appeared in America in the course of the Blancan, but whether they originated in the Old World or in the New is disputed. Walter W. Dalquest, of Wichita Falls, Texas, thinks that asses and hemiones on the one hand, and horses and zebras on the other, were separate lineages even before Blancan times and that most American horses except the zebras were in fact asinine. In this view, North America was the center of ass and hemione evolution—with some migrants to the Old World—while Eurasia was the playground of caballine horses.

BLANCAN SAFARI

The largest animals in the Blancan landscape, superficially resembling elephants, were the mastodons. By far the most common and wide-ranging seems to have been the "wonderful stegomastodon," *Stegomastodon mirificus*. It was not at all closely related to the famous American mastodon, which was also present in the Blancan, but of which I shall have more to say in a later chapter.

Mastodons, like men, came originally from Africa. Their earliest history, like that of man's early ancestors, is recorded in the deposits of Fayum, not far from Cairo, laid down by an ancestral Nile River more than 30 million years ago. Already at that early date, there is an incipient division into two distinct groups, which scientists call zygodonts and bunodonts; and the two lived and spread side by side. About 18 million years ago they reached Europe and Asia, and entered North America some four or five million years later. The bunodonts beat their zygodont cousins into South America.

But the histories of the two groups are very different. The zygodonts were cautious conservatives. Like their fellows they started out with tusklike incisors in the lower jaws as well as in the upper: they were four-tuskers. But, apart from a gradual reduction in the size of the lower tusks, until finally they were vestigial or lost, they did not change much. The American mastodon was the last of the zygodonts.

The bunodonts, on the other hand, were (biologically speaking) experimenters and inventors. In some of them the lower tusks became modified into weird-looking shovellike or spoonlike structures, or the jawbone itself became inordinately lengthened. Then, in the Pliocene, there appeared a general tendency to reduction of the lower tusks and jaws, in the bunodonts as in the zygodonts (and in the true elephants too). By Blancan times, all of the most peculiar bunodonts were extinct, with the exception of one rare four-tusked form.

The stegomastodons were among the last to carry on the heroic story of the bunodonts in North America. They had carried the evolutionary trend to a point where the lower tusks were completely lost and the skull and jaws were markedly shortened. This, in connection with the large, evenly curved upper tusks, must have given the wonderful stegomastodon a very elephantlike appearance in the flesh, although it was somewhat more stockily built than a modern elephant. With a shoulder height of about eight feet, this animal was close to a living Indian elephant in size. The American mastodon, largest of the zygodonts, may have stood one or two feet taller.

The first find of *Stegomastodon mirificus* was made by the great prospector F. V. Hayden in the 1850s, and the name was given by Joseph Leidy, often regarded as the founder of American vertebrate paleontology. The specimen came from the Loup Fork of the Platte River in Nebraska. At present, *Stegomastodon* is recorded as having lived in Arizona, California, Idaho, Kansas, Nebraska, and Texas, and it ranges in time from the early Blancan well into the succeeding Land Mammal Age, the Irvingtonian. Its extinction in Arizona is dated at 1.6 million years B.P.

Stegomastodons also spread into South America, where they survived much longer, up to the very end of the Pleistocene.

Next to the proboscideans, among the large and impressive animals of the Blancan, come the camels. If the idea of wild mustangs in North America is familiar to everybody, the notion of wild camels here may sound strange. But in fact North America is the original homeland of camels, as of horses, and even more so: up to the early Blancan, about 4 million years ago, *all* the camels in the world lived in North America. And their history, though not quite as long as that of the Equidae, is very well documented. Their fate, like that of the horses, has been to become extinct in their homelands, and survive in other areas. Today, true camels persist in the Old World, while the related llama group has survived in South America.

Both groups, the Camelini (true camels) and Lamini (llamas), were present in the Blancan. In addition, there is a third group, now extinct, the Camelopini or *Camelops* group. The division is an old one: the three lineages, camels, llamas, and camelopines, can be identified as far back as 15 million years ago.

The stateliest of Blancan camelids was *Titanotylopus*, which, as the name implies, was a large animal. It reached a shoulder height of almost 12 feet and walked on long, massive limbs. It had a single large hump, like today's dromedary. It was something of an oldtimer, having existed some millions of years before Blancan times, and one of the earlier species is thought to be ancestral to the Old World camels. So it belongs to the Camelini. This camel has not been found in Florida.

Compared with camels, llamas differ in the lack of dorsal humps, in the more highly domed shape of the skull (they have larger brains), and also in details of the teeth. The Blancan llama, *Hemiauchenia blancoensis*, was larger than living llamas; it ranged in time from the mid-Blancan to the early Irvingtonian, and in space from Alberta in the north to California, Arizona, and Texas in the south and east. An ancestral species is known from pre-Blancan times, and in the Irvingtonian, a daughter species, *Hemiauchenia macrocephala*, slightly smaller, was common.

The third camelid group, the Camelopini, is closely related to llamas but resembles the true camels in possessing a dorsal hump. It was present in the Blancan, but will be discussed in more detail in a later chapter. And the same holds for the pronghorn antelopes, whose history in the Blancan is imperfectly known.

Very few deer are known from the Blancan, perhaps because most deer are forest animals and many of the Blancan sites may represent habitats with open ground, gallery woods, and the like. The main exception is Santa Fe in northern Florida. Here, deer remains are plentiful, suggesting the presence of dense forest in the area. Surprisingly, the Santa Fe deer seem to be identical in every respect with the living white-tailed deer, *Odocoileus virginianus*. It is interesting to see that this species, still the most wide-ranging deer in America, is so old: for close upon three million years evolution seems to have passed it by.

This brings up an interesting point as regards evolutionary theory. In the traditional "gradualist" view, change is incessant: species evolve during their entire span of life and so gradually transform into new species. Now it has become increasingly clear that such is rarely the case. What happens is that species attain an equilibrium when well adapted to their environment and current mode of life, and then tend to remain static until some vital change occurs in the environment. The species may then respond in a positive manner by changing, or if this is not possible, in a negative manner by becoming extinct. In this view, which has been called the "punctuated equilibrium" theory by its proponents (Niles Eldredge and Stephen Jay Gould) evolution proceeds in comparatively rapid spurts in between periods of stasis. It is well illustrated by the white-tailed deer, which arose from an early Blancan ancestor (the Rexroad *Odocoileus brachyodontus*) and has then remained essentially unchanged. But this "punctuated" mode is only one extreme in a broad spectrum of evolutionary histories. At the other extreme is the "gradualistic" mode, of which an example will be discussed in the next chapter.

There are a few other species as ancient as the white-tailed deer. The oldest of them all seems to be the badger, *Taxidea taxus*, which has been traced to the early Blancan and back to a Hemphillian ancestor. In the late Blancan appear the following extant species:

opossum, *Didelphis virginiana*
masked shrew, *Sorex cinereus*
least shrew, *Cryptotis parva*
eastern mole, *Scalopus aquaticus*
long-tailed weasel, *Mustela frenata*
spotted skunk, *Spilogale putorius*
striped skunk, *Mephitis mephitis*
bobcat, *Lynx rufus*
beaver, *Castor canadensis*

Thus, among the many alien and exotic creatures of that ancient world, there were a few that are familiar to us. Perhaps these oldtimers can teach us something about survival!

Of course this has been a far from exhaustive list of the Blancan mammals. We have hardly looked at all at the small animals, among which are found forerunners of today's squirrels and marmots, pocket gophers, kangaroo rats, the various American "rats" and "mice," voles, muskrats and lemmings, jumping mice, hares and rabbits, and so on—many of them now known to have existed in large numbers. We shall look at some of them in the next chapter. The carnivores still remain to be discussed, as do some interesting invaders from the south.

Several Blancan sites have produced sizable avifaunas, which are of great interest, not only because they give us glimpses of the evolutionary history of American birds, but also because they give indication of the neighbouring habitat. Thus, the presence of pelicans, ducks, swans, and geese at Hagerman, Idaho, bears out the idea of a well-watered environment.

The "biggest in the world" department may look to the Blancan for the best fossils of one of the largest flying birds ever. Named *Teratornis incredibilis* by its awed discoverer, the Los Angeles paleornithologist Hildegarde Howard, it was originally found in a late Pleistocene cave in Nevada. It has since been identified from better (but still very scanty) material from the Anza Borrego Desert. Its size was about one-third larger than that of the related *Terratornis merriami* of Rancho La Brea, whose wing-spread was 12 feet (that of the living condor, greatest of raptors, is 10 feet). The wings of the incredible teratorn evidently measured 15 or 16 feet from tip to tip! (An even bigger teratorn has recently been found in South America.)

Among the reptiles of the Blancan—snakes, lizards, crocodilians, and others—the tortoises merit special mention, in particular the giant forms of the genus *Geochelone.* Related to the well-known elephant tortoises of the Galápagos Islands, these Blancan giants reached up to twice their size. They are common as far north as Kansas and Nebraska and so give independent evidence of a warm climate.

INVADERS FROM THE SOUTH

"It is odd that there should be *a* Latin American fauna," writes George Gaylord Simpson, "a broad unit that can be roughly designated by such a term as 'Latin American,' defined by human linguistics and culture." Yet this is true at the present day (although, as Simpson notes, its northern boundary does not coincide exactly with that of Mexico). For the origin of this situation we have to look far back into the geographic history of the Americas.

For most of the Tertiary, South America was an island continent, isolated by water barriers. The various primitive mammals which existed there at the beginning of the period followed their own evolutionary pathways, undisturbed by contacts with the outside world; and, as in other continents, there arose a balanced, dynamic animal world, wonderful in its richness and variety, and unique in its composition. Much of it is gone now; enough remains to give the Latin American fauna its unmistakable stamp.

Here, in isolation, archaic hoofed mammals gave rise to great stocks of peculiar ungulates. Some of them looked like nothing else on earth, but there are also many which paralleled unrelated animals in other parts of the world—pseudomastodons, pseudohorses, pseudocamels, pseudohippos, and so on. Evidently there were no true carnivores among the original inhabitants, for the carnivore niche was taken over by pouched mammals, marsupials, related to the opossum and to such Australian animals as the marsupial "wolves" and marsupial "cats." Among the native South Americans are also found the edentates, an order of mammals in which the teeth tend to dwindle, lose their enamel and, in some cases, become lost; it survives in today's armadillos, tree sloths, and anteaters.

Later on, a few other stocks entered South America, probably by waif dispersal across the water, on floating trees and the like. They most probably came from Central America, which at that time was connected with North America. The strait separating the two continents was across the northwestern corner of South America, parting Colombia from Panama. (A trans-Atlantic route has also been considered; the South Atlantic of the early Tertiary was much narrower than now.) These invaders gave rise to the South American monkeys and the caviomorph rodents, surviving today in such forms as guinea pigs, chinchillas, water hogs, and porcupines.

At about the beginning of the Blancan, movements in the earth's crust resulted in the emergence of a land bridge from South America to Central America and further northward. Thus began what has been called the Great American Interchange. A host of land mammals from North America invaded the south, and as a result many of the native South American forms died out. But many of the southerners were hardy enough to stand their ground, and not a few migrated northward to populate North America. Some of them were destined to become important and striking elements in the North American animal world of the Ice Age.

The interchange actually started several million years earlier, but these early migrants are so few that it is thought they spread across a water barrier. At this time, about eight million years ago, two South American groups of ground sloths (mylodons and megalonychids) appeared in North America, while some raccoonlike animals crossed in the other direction. Waif dispersal is somewhat unlikely for the ground sloths, unless the early forms were very small, but at least the megalonychid sloths may have been powerful swimmers, as indicated by their colonization of the Antilles during the Ice Age.

The next migration wave came in early to mid-Blancan time, about 3 million years ago. By then an overland route must have been in existence, for the impact, especially on the fauna of South America, was great. Still other bursts of migration came near the end of the Blancan, and later on.

Although a few of the native South American ungulates and monkeys extended their ranges into Central America, they did not reach the United States, perhaps because the climate was unsuitable or because they could not compete with their Northern rivals. In fact those forms that did succeed were of a kind that had no close counterparts in the North American fauna. Among them we find the opossums; the armadillos and the related huge, armored glyptodonts; the ground sloths; the water hogs; and the porcupines. Except for the opossum, which of course is also unique in many respects, the successful invaders listed here were plant-eaters utterly different from anything that existed in North America, and so probably not in competition with the native fauna. Rather, they were able to occupy ecological niches that up to then had been empty. We shall deal with many of these creatures in later chapters.

Titanis, the thunderbird of Florida, was a great, rapacious, flightless bird.

But there is one invader from the south which does not fall in the same category. There are but few remains of it, and most come from the Santa Fe River site in North Florida. Here, Pierce Brodkorb found the bones of an immense flightless bird which he named *Titanis walleri*. It was about the size of an ostrich but evidently very different from that basically inoffensive plant-eater. It belongs to an extinct South American family, the Phorusracidae, giant seriemas or thunderbirds.

A closely related family, the Cariamidae, with two species—the seriema and the chunga—is still extant in South America. These are birds of the open plains, very fast and graceful runners that only occasionally take wing. Although related to cranes, they have a curved beak resembling that of a bird of prey. They hunt insects, frogs, and reptiles, especially snakes, and for this reason are highly esteemed by the people. We may imagine the history of the thunderbirds beginning with something like a seriema, gradually specializing in bigger game, growing larger, and losing the power of flight.

Such great rapacious ground-birds seem to be able to flourish only in areas where there are no true Carnivora, as was the case in South America, where the only competition came from marsupials. Still, in the Blancan they temporarily extended their range to Florida. *Titanis* may have traveled there by way of the Gulf Coast, and it must have been a spectacular element in the Blancan scene at Santa Fe.

In the long run, Waller's thunderbird probably was crowded out by carnivorous mammals competing for the same prey, and perhaps even preying on *Titanis* itself. There were indeed plenty of carnivores present, from small weasellike creatures to lion-sized sabertoothed cats. None of the South American marsupial carnivores is known to have invaded North America, and so the flesh-eating mammals of the Blancan are a mixture of indigenous forms and invaders from the Old World. The latter have furnished the zoogeographers with some real surprises, and I shall begin my account with one of these.

WITNESS OF THE CHASM

There is a bone that is one of many from a fissure filling in Permian limestones near Anita in the Plateau Region of northern Arizona. Among the animals are hare, woodrat, an extinct long-nosed peccary, and a large horse. But this bone is the lower jaw of a carnivorous mammal, perhaps the one that denned in the fissure and brought in the prey.

It's a sorry specimen. All of the teeth and much of the jaw have been lost, and all that remain are the front and hind end, united at the lower border, and some roots and fragments of teeth. Holding it in my hand I can estimate that ninety-nine out of a hundred paleontologists would file it away as "unidentified carnivore."

But it so happened that the man to whom it was entrusted was one in a hundred, and thus starts an investigation as complex as many a detective story.

The man was Oliver P. Hay, at that time the leading Pleistocene paleontologist of the United States. The time was about 1920, the place the Smithsonian Institution. The bones from Anita are still there.

Poring over the sad remains, analyzing the scraps of evidence, he eliminated one possibility after the other. Not a cat. Not a dog. Certainly not a bear. And so on. And at last there was only one possibility left. Of all known carnivores it could only be a hyena.

But no one had ever heard of hyenas in North America!

Fossil hyenas are a commonplace in the Old World, where their abundant record goes well back into the Tertiary. But an American hyena? On *that* kind of evidence?

Undaunted, Hay published his results in 1921. He named the animal, rather grandly, *Chasmaporthetes ossifragus*. *Chasmaporthetes* means "the one who saw the chasm open up"—Hay was thinking of the Grand Canyon! *Ossifragus* means bone-eating: hyenas are known to chew bones.

It may be well to digress for a moment on the scientific names of animals. The Swedish eighteenth-century naturalist Carl von Linné (Linnaeus) established the binomial system of nomenclature, in which the name of a species consists of two parts. The first part, which is capitalized, is the name of a genus (which may comprise several related species). The second part, which is not capitalized, is the "trivial" name. (It is sometimes, erroneously, called the "species name," but in fact the species name is the combination of generic and trivial names.) According to the International Rules of Nomenclature, the earliest name is the valid one, and later names for the same animal are set aside as synonyms. The name will remain valid even though its literal meaning turns out to be incorrect. It's just a tag, and so *Chasmaporthetes* remains *ossifragus* even though we now know that it was not a bone-eater.

The sensational news that there had been wild hyenas in North America, then, was given to the world in 1921, and did anyone bat an eye? It seems that the profession shrugged its collective shoulders. Hay had overreached himself, and it was better to forget the whole thing.

Seventeen years later a well-preserved jaw turned up in the Cita Canyon fauna of the Texas Panhandle. It was described by Professor Ruben A. Stirton of Berkeley, together with Wayne C. Christian, and they came to the conclusion that it was a hyena. But so completely had Hay's paper been forgotten that they, in the honest belief that this was a new thing, coined a new name for the creature. Later, of course, they recognized Hay's priority and referred their animal to *Chasmaporthetes.*

Here was evidence of a different caliber. All the teeth were preserved. But although hyenalike, they were not of the bone-crushing type. They were slender and sharp-edged, and rather resembled those of a big cat. Only there were too many of them. The lower jaw of the cat has two premolars and one molar. *Chasmaporthetes* has three well-developed premolars and one molar, just like the hyenas. Hay's analysis was confirmed, yet many students remained unconvinced.

Meanwhile, things were happening in Europe. In 1914, the Italian student D. Del Campana had described a hyena from the Val d'Arno beds in Tuscany under the name *Lycyaena lunensis. Lycyaena* is a well-known late Miocene genus of Old World hyaenids: these animals lived about 7–14 million years ago. They lacked the bone-crushing specialization of modern hyenas and had slender, catlike teeth. But Del Campana's animal was much more recent than the Miocene. Later discoveries have shown that it lived in Europe roughly between 3.5 and 1 million years B.P.

Next, the Swiss paleontologist Samuel Schaub, in 1941, collected all the available data on Del Campana's beast and was able to demonstrate some truly astonishing characters. It must have been a highly predaceous carnivore with very long and slender, almost cheetahlike legs. Also, it was much more advanced in its dentition than any *Lycyaena,* and so he created a new genus for it, *Euryboas.* (There are some added complications which will be skipped here.) He did point to the resemblance to Stirton's and Christian's Texan beast, and concluded that that, too, was a hyena.

So now we have *Chasmaporthetes ossifragus* in America, and *Euryboas lunensis* in Europe. But the story goes on.

Could the two be related? What we need is a geographic link-up. Actually, part of such a link-up had been in existence since 1932, when the Russian paleontologist P. Khomenko described a near-complete skeleton of a young hyena from early Pliocene deposits in the Moldavian Soviet Republic. We can now recognize this animal, which Khomenko called *Hyaena borissiaki,* as an early (about 4–5 million years old) *Chasmaporthetes.* But the connection went unnoticed until the Swiss paleontologist Gérard de Beaumont pointed out the resemblance in 1967, and even then few took notice because the remark was an aside in a treatise dealing with some other animals.

Next come discoveries of *Chasmaporthetes* in Transbaikalia and in Mongolia, by Marina Sotnikova of Moscow. Then, in 1977, Henry Galiano and David Fraily of the American Museum of Natural History extended its range to China. So we have gone a long way toward forging a chain from western Europe to eastern Asia and North America.

Meanwhile, new records of American hyenas have come to light: from Mt. Blanco, Texas; from Comosí, Arizona; and from two Mexican sites, Goleta in Michoacán and Miñaca Mesa in Chihuahua, among others.

But the clinching evidence comes from the site of Inglis in Citrus County, Florida. While the sites mentioned above are Blancan in age, Inglis is thought to be slightly later: *Chasmaporthetes* survived (at least in Florida) after the end of the Blancan. Here, for the first time in America, S. David Webb and his student Jean Klein found limb bones together with the teeth and jaws of the American hyena—and they are quite similar to the European ones that Schaub described. So it seems we have come full circle.

Such is the story that begins with a singularly unprepossessing scrap of bone from Arizona, and ends with a nearly worldwide population of this remarkable cheetah-hyena (it has been found in Africa, too). In America it probably arrived at the beginning of the Blancan, like so many other migrants across the Bering land bridge. I have told the story in some detail to give an idea of the sometimes remarkably tortuous routes by which a synthesis may be built up in paleontology, and the way in which the evidence can be pieced together.

"RED IN TOOTH AND CLAW"

Flesh-eaters rely on many different strategies to take their prey. Some, like the leopard, may lie in ambush, Others, like the fox, rely on stalking to get close. Some of the most intelligent, like the wolf, the lion, and the spotted hyena, use flock tactics. There are also a few that are able to outrun anything on two or four legs, and catch their prey in an overwhelming burst of speed. It is just possible that *Chasmaporthetes* was one of these; what is certain is that the cheetah is a superb living example.

It might be thought that there was no room for more than one predator of this kind. Yet we find, oddly, *Chasmaporthetes* and cheetahs associated all over the Old World in the late Pliocene and early Pleistocene. And so it is in North America too.

The story of this discovery is about as complicated as that of *Chasmaporthetes*, and some of the same characters (for instance, the Swiss paleontologist Samuel Schaub) figure in it. However, it is only in the last few years that Don Adams has been able to conclusively show that the American cats in question were in fact cheetahs. He was met by the same incredulity as Hay (e.g., by referees of scientific journals) but I think he has proved his point. There were indeed cheetahs in North America, in the Blancan and throughout the Pleistocene.

The Blancan cheetah, *Miracinonyx studeri*, was a giant form fully as large as its European counterpart and close relative, *Acinonyx pardinensis* (which was pieced together by Schaub). The forest-savanna of the Great Plains, where it has been found—at Cita Canyon—may have been an ideal setting for this predator.

The living cats show a wide spectrum of adaptive types, many of which were also present in the Blancan. One kind, however, is now extinct: those in which the upper canine teeth were greatly enlarged and used for stabbing rather than biting. They are the so-called saber-toothed "tigers" or sabertooths. They are not specially related to tigers, or to any other living cats, but belong to an extinct branch of felids. Most of them probably fed on large, slow-moving prey, and so, contrary to cheetahs, they were built for power rather than speed.

The biggest of the Blancan sabertooths, *Ischyrosmilus*, the size of a lion, became extinct at the end of the Age. A smaller form, but one destined for a remarkable future history, was the Western dirktooth, *Megantereon hesperus*. It seems to have originated in the Old World, for dirktooths are found in Eurasia and Africa and their story goes back to the late Miocene. About the size of a puma, the dirktooth differed from the puma in its much heavier build and above all by the presence of extremely long fangs in the upper jaw, protected by a peculiar flange at the chin of the lower jaw. As in other members of the saber-toothed tribe, the jaw joint and musculature were much modified to allow a gape wide enough to clear the points of the fangs. I shall have more to say about this tribe in a later chapter when we come to *Megantereon*'s Pleistocene descendant, the great sabertooth *Smilodon*. But it may be noted that the dirktooth (like the descendant) seems to have been a very successful animal, present in forest as well as savanna biotopes, and ranging from Idaho to California and Florida. In the Old World, too, its ubiquity is remarkable.

But what about the role of the hyena? Cadavers of large animals are dealt with by many kinds of scavengers. The hyena's specialty is feeding upon those parts which other mammals are unable to deal with, such as the toughest skin and the nutritious parts of bone. If left alone, these too will be ultimately handled by a host of small organisms which break down the edible stuff piecemeal. But there is always room for a hyena, and, as we have seen, this was not the niche of *Chasmaporthetes*.

The part of the hyena was played in the Blancan by a bone-eating dog: *Borophagus diversidens*. (It is one of the few animals to be unpopular several million years after death. Paleontologists do not like to have their bones broken.)

The history of the dog family is more varied than we might imagine if we just look at its living members—wolves, dogs, foxes—which, in spite of differences in size and coloring, have so much in common: there is no mistaking a canine face. But in the past there have been other types of dogs, and some must have looked very different in the flesh: catlike dogs, bearlike dogs, weasellike dogs, hyenalike dogs. Almost all died out before the Blancan: the hyenalike dog was the sole exception.

Borophagus was a wolf-sized creature, but in appearance it was very different from the wolf. It was comparatively short-legged, and the head was very big, with short jaws and almost fantastically massive teeth. Some of these teeth are so like those of true hyenas that an isolated specimen could very easily be mistaken for one. Still, the total ensemble is nonetheless that of a dog, although weirdly transformed. That *Borophagus* did in fact smash bones there can be no doubt. *Borophagus*-chewed bones are common. And the hunting seems to have been good, for remains of the bone-eating dog have been found at many sites from Florida to Idaho. Extinction came at about the end of the Blancan, but in southern Alberta *Borophagus* survived long enough to confront immigrant mammoths in the early Irvingtonian.

It would carry us too far afield to take a detailed look at all the Blancan carnivores, which make up a wonderfully varied roster. There are some startling elements, for example a panda from the state of Washington. It is not the giant panda (which is common enough in Chinese caves, but unknown in the New World) but a form related to the red panda of the Himalayan forests. The fossil species also occurs in England. As that was where it was found first, it is called the English panda, *Parailurus anglicus*. Clearly, once upon a time, the English panda was also a Russian, Siberian, Beringian, and American panda.

In Idaho, on the other hand, we find relatives—perhaps ancestors—of the South American mustelids called grisons and tayras, large martenlike animals. Like the panda, they came from Eurasia. We can see glimpses of long stories of migration, only half-unraveled, and see how many animals have trodden these paths that man, millions of years later, was to follow.

At the end of the path, however, we might pause to look at one dog in particular. It looks more familiar than *Borophagus*, and for good reason. It is a coyote. I have spoken of exotic-looking dogs, but here we have a doglike dog. And, like the horse and the camel, it is an American aborigine. The ancestry of the central dog genus, *Canis*, can be traced back in North America for thirty million years. In the Blancan this line had progressed far enough to produce an animal essentially like the modern coyote. There are still some differences, mainly in the limb proportions—evidently it was not quite as fleet of foot as the living species. The Blancan coyote is called *Canis lepophagus*, meaning "rabbit-eating dog": probably a very suitable appellation. There can be little doubt that it was directly ancestral to the living species, *Canis latrans*. But again, there is a link with the Old World. An animal which lived in Europe about 1.5 million years ago, long thought to be a jackal, has turned out to be a coyote. So this little "wolf of the prairie," as it is called in Europe, did get around too.

Soon there was to be a new burst of migration across the northern lands of Beringia, and that one marks the transition from the Blancan to the Irvingtonian: 1.9 million years ago.

III

The Coming of the Cold

THE IRVINGTONIAN AGE

THE IRVINGTONIAN LAND-MAMMAL AGE takes its name from a fossil-bearing site near the town of Irvington, in the southern San Francisco Bay region of California. Donald E. Savage of Berkeley, in his 1951 study of this locality (which yielded the remains of various fish, frogs, toads, turtles, a Canada goose, and 21 species of mammals), pointed out the key characters of the fauna. Two are of special importance.

In the first place, there are mammoths at this site. There were no mammoths in the Blancan, so this fauna must be younger, dating from a time when the mammoths had already entered North America from their original homelands in the Old World.

In the second place, there are no bison. Of course, this can be disputed as negative evidence. There might have been bison which did not happen to be sampled in the fossil record. On the other hand, experience has shown that bison do tend to leave a record—they are among the most commonly found fossils in late Pleistocene deposits. There is thus a suggestion that the Irvington fauna antedates the entrance of bison in America—and they, like the mammoths, came from the Old World. On this basis, Savage characterized the Irvingtonian Age: the interval between the immigration of mammoths and that of bison.

Further studies have borne out Don Savage's concepts and shown that the Irvingtonian is indeed a well-characterized stage in the evolution of the North American animal world. Its beginning can be pinpointed with some precision because it coincides very closely with a paleomagnetic subchron (a brief reversal of polarity) known to have occurred around 1.9 million years ago. Exactly when the Irvingtonian ended is a more difficult matter. This happened well within the last paleomagnetic chron, the Brunhes, so all that can be said on this score is that it occurred later than 700,000 B.P. when the Brunhes chron started. Other evidence indicates a substantially later date: for instance, a late but by no means the latest Irvingtonian fauna from Cudahy, Kansas, is reliably dated at 600,000 B.P. A rough guess at the time of transition from the Irvingtonian to the late Pleistocene age, the Rancholabrean, is of the order of 400,000 years. So the Irvingtonian lasted about 1.5 million years, or about the same length of time as the Blancan.

Within the span of the Irvingtonian fall at least two of the great continental glaciations, as suggested by the chronology of deep-sea cores. The first, or Nebraskan, commenced about 1.5 million years B.P. and was followed by a warm interval, the Aftonian interglacial. Then, seemingly about 900,000 B.P., came the Kansan glaciation, succeeded by the Yarmouthian interglacial. The third glaciation, the Illinoian, probably started about 600,000 B.P.; and so its early part, at any rate, falls within the Irvingtonian.

Much remains to be done before we have a reliable chronology. Still, one thing is evident: the Irvingtonian was the time when the Ice Age came to America. Glaciations and interglacials unrolled in their full amplitude, and the Pleistocene was under way.

The story of the Irvingtonian Age, the time span between the entrance of the mammoth and that of the bison, is being pieced together from many sites, covering an even greater area than that of the Blancan. There are now important localities as far to the northeast as Pennsylvania, and most of these are caves.

The cave, or fissure, or sinkhole, is often an ideal environment for the preservation of fossil bones. Many animals, especially carnivores and birds of prey, use caves or fissures for dens. They may die there and leave their bones on the cave floor, together with remains of their prey. Owls are especially productive: they regurgitate the undigestible remains in the form of pellets which are stockpiled beneath their roosting places. Bats, of course, very frequently inhabit caves, and their guano contributes to the sediment on the floor, where their own bones may be imbedded. Other animals may fall into sinkholes; they may have been killed or maimed, or simply unable to get out.

So far, four fossiliferous caves of Irvingtonian age have been studied within a belt ranging from the Ozarks to the Appalachian Mountains. Probably the oldest, and also the first to become known, is a cave at Port Kennedy in Montgomery County, Pennsylvania. It was studied by Edward D. Cope and was the subject of his last great work, published posthumously in 1899. More than forty mammalian species are known from here. Among the most notable are the earliest record of wolverine and black bear in North America, and a transitional form between the Blancan dirktooth *Megantereon* and the later *Smilodon*. Two other caves in the same general region, Trout Cave in the central Appalachians near Franklin, West Virginia, and Cumberland Cave in Alle-

Earliest wolverine in America may be circumpolar *Gulo schlosseri*, found in Port Kennedy and Cumberland Caves. It was somewhat smaller than the living species.

gany County, Maryland, date from a later phase of the Irvingtonian. Cumberland Cave, which was exposed by a railroad cut, is remarkable for the large number of black bears—perhaps fifty or more—which left their bones there. Still another late Irvingtonian cave, Conard Fissure, is situated in the Ozark Mountains, close to Buffalo River in Arkansas. All of these have yielded great numbers of fossils, on a par with the Port Kennedy Cave.

In Florida, two productive Irvingtonian sites of this type have been found. A cave by Inglis, Citrus County, sampled the inhabitants of the surrounding coastal savanna, probably in the very earliest Irvingtonian, for *Chasmaporthetes* was still in existence here. And a filled-in sinkhole at Coleman in Sumter County has yielded remains of small mammals preyed on by owls, as well as the bones of large animals that fell into the sink—including several enormous jaguars. That was in the late Irvingtonian, and the landscape appears to have been more open, with a dryer climate, than it is today. It is thought that glacial conditions then prevailed in the north, so the Coleman fauna may date from the early Illinoian.

Other known sites of Irvingtonian date are in river or lake sediments, and in some areas they follow in direct stratigraphic superposition upon the Blancan deposits. The sequence in the Anza Borrego Desert and in the San Pedro Valley, in California and Arizona respectively, continues well into the Irvingtonian, and the magnetic stratigraphy gives unique opportunities for dating the transition. For instance, in the San Pedro Valley, a very rich earliest Irvingtonian fauna at Curtis Ranch dates from the beginning of an important magnetic subchron called Olduvai (after the place in Africa where it was first identified), and was known to occur in the interval between 1,860,000 and 1,710,000 B.P. It is thrilling to reflect that, with the help of magnetic measurements, we are able to say that the Curtis Ranch animals lived at exactly the same time as certain proto-human beings in what is now Olduvai Gorge in East Africa. Their descendants were destined to meet on American soil, nearly two million years later.

Giant tortoises, like this living species from Aldabra, are common in Blancan and Irvingtonian faunas. Their presence indicates mild winters.

In the Great Plains, the climate of the Irvingtonian was dryer than in the preceding age, and the Blancan forest-savannas were superseded by grassland with lake-woods and gallery forests. The swing between cold and warm, dry and moist, led to repeated cycles of sedimentation and erosion and this has resulted in the formation of a series of river-laid terraces which can often be followed over great distances and are very useful in correlation. The terraces, and their contained fossils, have been the subject of decades of study, for instance by C. Bertrand Schultz and his associates at Lincoln, Nebraska.

Among the many important sites in Nebraska, Kansas, Oklahoma, and Texas, I will choose one as an example of how the geological and paleontological analysis may result in a surprisingly detailed picture of the environment of the past. The site is Slaton Quarry near Lubbock, Texas, which was studied by Walter W. Dalquest of Wichita, Texas, in the 1960s.

The bones come from the bed of an ancient lake which had formed in a deep depression in the underlying Pliocene rocks. We know the lake was fairly big because there were great alligators in it. There were also several islands of varying size. The bone-bearing clays accumulated near the northern margin of the lake in shallow, vegetation-choked waters. As time went on, conditions became slowly dryer and the final filling of the bed took place in a semiarid climate as suggested by the increasing amount of sand. There were lots of water rats in the lake. At least in the earlier stages, when the climate was still fairly moist, there must have been trees and thickets along the shores, for deer and armadillos were common. The presence of armadillos (and also alligators and box turtles), incidentally, shows that the climate was mild, with frost-free winters. But the area around the lake must have been grassland, for there are no large browsing mammals but instead an abundance of grazers like camels, pronghorn antelopes, horses, and mammoths. Bison would certainly have been expected here if present, but there are none; so Dalquest considers it a pre-*Bison* fauna. But the date

Pikas are related to hares and rabbits, but have short ears and legs. They make hay—collecting grasses and herbs, and stack them in the sun to dry. Their record goes well back into the Irvingtonian.

is probably late Irvingtonian, for many of the small mammals—ground squirrels, pocket gophers, pocket mice, kangaroo rats, harvest mice, and woodrats—belong to modern species.

In Pleistocene deposits, we often find associated remains of animals which are now widely separated geographically. Thus, at Slaton, there are water rats—which are now found in Florida and Georgia—together with the vagrant shrew, a western and northwestern species. In such cases we may suspect that Pleistocene conditions as regards climate and other environmental factors, differed from those obtaining anywhere today.

In the northwest, as in the northeast, the Irvingtonian record is greatly extended, compared with the Blancan. A very interesting site, still being studied by Charles S. ("Rufus") Churcher, A. MacS. Stalker, and their associates at Toronto, is Wellsch Valley, which is located in southern Saskatchewan about 25 miles north of Swift Current. Buried beneath a series of at least four glacial tills, the bone-bed—a pond deposit—has yielded a unique association of mammoth and the Blancan bone-eating dog, *Borophagus.* It suggests that we are here in the very earliest phase of the Irvingtonian. Near Medicine Hat in Alberta, there is a long series of glacial and interglacial deposits in stratigraphic superposition, extending from the middle Irvingtonian to the late Pleistocene. And, finally, a site at Cape Deceit on the southern shore of Kotzebue Sound, Alaska, has produced a glimpse of Irvingtonian life in the treeless lowland tundra of Beringia. So we have a wide spectrum of Irvingtonian environments and faunas, from the arctic tundra to the subtropical forests and savannas of Florida.

EARLY MAMMOTHS AND MUSKOXEN

Mammoths belong to the elephant family, which arose in Africa in the late Miocene and gave rise to three great lineages. Two are still in existence today: the African elephants *(Loxodonta)* and the Indian ones *(Elephas).* The third, and extinct, genus is that of the mammoths, *Mammuthus.* Contrary to *Loxodonta* and *Elephas, Mam-*

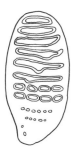

Evolution of mammoth cheek teeth features a progressive increase in the number of plates formed by the transverse ridges, embedded in cement. Left, southern mammoth *Mammuthus meridionalis)*, early Irvingtonian; center, Columbian mammoth *(M. columbi)*, late Irvingtonian and early Rancholabrean; right, woolly mammoth *(M. primigenius)*, Rancholabrean (the hooked shape of the hind end is pathological and not typical). Both *M. columbi* and *M. primigenius* were derived from *M. meridionalis*, along distinct lines of descent.

muthus became adapted to a cold climate, and this is evidently the reason why this was the only elephant genus which succeeded in crossing Beringia into the New World.

There are various differences between mammoths and other elephants—for instance, the mammoth skull is higher and more foreshortened than that of elephants. Also, although the cheek teeth in early species of *Mammuthus* and *Elephas* are similar, the number of transverse plates in advanced mammoths became much greater than in elephants. This is an adaptation for feeding on the very tough grasses of the steppe and tundra. But the most striking difference is the shape of the tusks. While those of other elephants are only slightly curved, or even nearly straight, the mammoth tusks were quite extravagantly curved, with a spiral twist, and mostly extremely long. This led in some cases to the points of the tusks crossing each other in front of the head.

When speaking of mammoths, we tend perhaps to think of the famous frozen carcasses of late Pleistocene specimens found in the tundra of Siberia and Alaska. These are woolly mammoths, *Mammuthus primigenius*, and will be treated in the next chapter. But the mammoths that entered North America in the Irvingtonian belonged to a much earlier species, the southern mammoth *Mammuthus meridionalis*. Present in the Old World since the Pliocene, this was still a rather primitive species. Its cheek teeth are readily identifiable from the low number of transverse plates, which form characteristic lozenge-shaped figures on the wearing surface. The tusks of the southern mammoths were straighter than those of the later forms, and in the flesh it would probably have resembled a modern Indian elephant, lacking the peaked head and sloping back so typical of advanced mammoths. It is also probable that the southern mammoth did not possess such a well-developed woolly coat of fur as the late Pleistocene northern species. Its size, however, was impressive: the shoulder height was of the order of 12 feet.

So far, there is no certain information on the exact time of arrival of the southern mammoth in North America. The record from Wellsch Valley may well be the oldest; as noted above, this mammoth is associated with *Borophagus*, a Blancan holdover. One find from Bruneau in Idaho is dated at 1,360,000 B.P., and in the San Pedro

Valley the earliest mammoth occurs at about the same time. The time of migration through Beringia may have been substantially earlier, perhaps as early as 1,900,000 B.P., when several species are known to have crossed. It might be speculated that the species, which had to adapt to northern conditions in order to populate Beringia, took a long time to readapt to a warmer climate and spread southward. Perhaps the push given to it by the Nebraskan glaciation may have speeded its southward migration. But such speculations may be nullified by future finds.

Still, though details of the migration story remain to be worked out, it seems that in the Aftonian interglacial, perhaps about one million years ago, *Mammuthus meridionalis* ranged from Europe to North America, and as far south as southern California, southern Arizona, and northern Florida. As in the case of other Holarctic species, it was partitioned by the Beringian bottleneck, where the gene-flow between the continents was restricted and intermittently stopped by interglacial flooding. Genetic continuity might prevail for many millennia, or even hundreds of millennia, but in the long run a divergence between the mammoths of the Old World and those of the New would be almost inevitable. Vincent J. Maglio, foremost student of elephant phylogeny, thinks that different lineages of descent were established on the continents of Eurasia and North America at the stage of *Mammuthus meridionalis* or soon afterward.

In the Old World, mammoths continued to evolve in adaptation to an arctic environment: the temperate and warm zones had already been claimed by the true elephants, *Elephas* and *Loxodonta*. In America, on the other hand, where no such competition existed, the mammoths were able to invade this niche. It might be thought that they had to do so in competition with mastodons, but that was probably not the case. Mastodons were browsers and so were unable to exist in the treeless plains which expanded greatly in the dry Irvingtonian climate. The prairies became the homeland of the mammoths.

Vince Maglio suggests that the southern mammoth gave rise to the Columbian (or imperial) mammoth of the later Irvingtonian. Various names have been applied to these animals, which survived well into Rancholabrean times, but Vince suggests they can all be united into a single species, *Mammuthus columbi*. It differs from its predecessor in the higher skull and more sloping back, and in a gradual increase of the number of plate-ridges on the molars, which indicates greater grazing efficiency. In this way, probably, the durability of the teeth was increased, so that individual life was lengthened; it is easy to see how this would be useful, and favored by natural selection.

Perhaps the most striking difference is in the shape of the tusks, for it is at the *Mammothus columbi* stage that they reach the size and degree of twisting which caused the tips to turn inward and even to cross each other, especially in large males. This is a characteristic feature of the American-lineage mammoths in the late Pleistocene, and will be discussed in a later chapter.

Black-tailed jack rabbit *(Lepus californicus)* may be the oldest species of the hare genus *Lepus*. Early finds date back 1.9 million years, to the Blancan-Irvingtonian transition.

In the Columbian mammoth, the American line attained the culmination of its size, and some of these animals were certainly among the largest elephants ever to exist. A specimen found in Lincoln County, Nebraska, and now on display in the University of Nebraska State Museum, attained a shoulder height of more than 13 feet. It is interesting to see that in the Old World, too, mammoths attained maximum size at this time. The skeleton of a mammoth that lived 700,000 years ago in Germany, now in the natural history museum of Mainz, belonged to an even bigger individual, but is unfortunately incomplete.

Besides the mammoth, several other migrants feature importantly in the definition of the Irvingtonian. One is the hare genus, *Lepus,* which appears simultaneously in North America and Eurasia at 1.9 million years B.P. Unfortunately, authorities do not agree on whether it originated in the Old World or in the New; possible ancestors are known in both areas. Most probably of Eurasian origin, on the other hand, is the jaguar—surprising as this may seem, for an animal that now is so typical of the "Latin American" fauna. But ancestral jaguars are known in Eurasia, and survived there up to about 600,000 B.P. In America, this cat pops up for the first time in the Curtis Ranch fauna, close to 1.9 million years ago.

If the cases of hares and jaguars are arguable, there can be no doubt that the muskoxen are of Eurasian origin. They belong to the bovid family, together with such animals as cattle, bison, sheep, goats, and Old World antelopes. No bovid whatever has been found in pre-Irvingtonian deposits in America. The first bovids here appear at the beginning of the Irvingtonian, and they are now thought to belong to the muskox group.

There were in fact two quite distinct forms, and it is only in recent years that C. R. (Dick) Harington, at the National Museum of Canada, Ottawa, has shown that they were related to the living muskox. The first to appear was the shrub ox, *Euceratherium collinum.* This was a large bovid, about four-fifths the size of a modern bison, and with big, upright horns very different from the flattened, drooping horns of the muskox. They rose steeply from the crown of the head, curved outward and forward, and then upward at the tip. The structure of the neck joint and vertebrae suggests that these animals used their horns in ramming bouts as do modern sheep and goats. Claude Hibbard thought they probably inhabited hilly country, much like today's wild sheep.

The shrub ox is first met with in the earliest Irvingtonian, but survived to the end of the Pleistocene. No closely related form is known from Eurasia, so its ancestor remains to be discovered.

The second species is Soergel's ox, *Soergelia mayfieldi*. Its generic name commemorates the outstanding German paleontologist and Pleistocene specialist, Wolfgang Soergel of Freiburg. First discovered in mid-Pleistocene deposits in Thuringia, Germany, it was studied by the Swiss Samuel Schaub (whose name we have met already in connection with *Chasmaporthetes*), who in 1951 suggested it might be an aberrant member of the goat-and-sheep tribe. The trail of the mysterious *Soergelia* was then picked up by others, notably the Soviet paleontologist Andrei Sher in 1975. He charted additional finds in Europe, from Czechoslovakia and Rumania; in western Siberia, in the valleys of the river Ob and two of its tributaries; and in eastern Siberia, in the Yana and Kolyma river valleys. So now we have *Soergelia* ranging all over Eurasia in the middle Pleistocene.

As early as 1915, Edward L. Troxell, then at Yale University, found the first American *Soergelia*. He studied a large fauna from Rock Creek, a site in Briscoe County, Texas. Here, besides the southern mammoth, a large horse, the great tortoise *Geochelone*, and various other animals, were found the remains of a bovid which seemed to be allied to the shrub ox. He tentatively placed it in the same genus as a distinct species, *Euceratherium mayfieldi*. The size of a steer, this heavily built animal had shorter horns than the shrub ox, and they were directed outwards rather than upwards, and slightly downcurved. Dick Harington has pointed out that this animal is in reality *Soergelia*—a genus unknown at the time of Troxell's work—and that *Soergelia* is intermediate between the high-horned shrub ox and the low-horned muskox. So they all seem to be related. Moreover, Dick also has *Soergelia* in Canada, at a locality by the Old Crow River in the Yukon.

The Irvingtonian seems to have been a time of almost explosive evolution of the muskox tribe. There is a third form, called *Praeovibos priscus*, in which the horns droop as in the true muskox; it differed from the modern form in being rangier in build, with long legs. Its remains have been found in a northern belt extending from England and the Pyrenees to the Kolyma River in eastern Siberia, and into the American part of Beringia, where it occurs near Fairbanks, Alaska, and at the Old Crow River in the Yukon. So far, there is no evidence that this animal found its way into America south of the ice.

In addition to all these remarkable creatures, true muskoxen of modern type existed in Europe in the middle Pleistocene, but there is no evidence as yet of their presence in North America during the Irvingtonian.

MUSKRATS AND EVOLUTION

Given a continent like North America, with all its wonderfully varied environments, and its connections to other land masses, the world of living beings that inhabit it may be seen as a dynamic system. The number of species may remain approximately constant in the long run, yet the actors change as new species take the places of old.

There are three factors of change. One is extinction. It need not be dramatic. Perhaps a species finds the going a little tough: the animals do get on but not quite so well as they used to. So the numbers thin out, generation after generation. The point of no return is reached when the population is so sparse that males and females do not easily find each other. In most cases it is impossible to tell just *why* the extinction occurred. Competition with new species—immigrants, or locally evolved—may have been a decisive factor in some cases, climatic change in others.

This fate befell many of the Blancan species. The American zebra and the tiny gazelle-horse did not survive the end of the Blancan; the hunting hyena *Chasmaporthetes* and the bone-eating dog *Borophagus* barely lasted into the early Irvingtonian. Gone before the end of the Blancan, too, was the great sabertooth *Ischyrosmilus*. The stegomastodons lasted into the Irvingtonian but vanished about 1.6 million years B.P. A highly varied group of primitive hares and rabbits vanished completely and were replaced by the modern genera *Lepus* and *Sylvilagus*.

The second factor is immigration. We have seen already how immigration from the Old World affected the Irvingtonian fauna. There were a few immigrants from the south, too. Capybaras and porcupines entered North America about 2 million years ago; somewhat later, perhaps 1.8 million years B.P., appeared new types of ground sloths *(Eremotherium* and *Nothrotheriops)* and a giant armadillo *(Holmesina)*. They are treated in later chapters.

The third factor is evolution. A species may vanish, not because it becomes extinct but because it evolves into a new species (or into several daughter species), for instance the rise of modern coyotes from the Blancan *Canis lepophagus*. The Irvingtonian animal world was rejuvenated in this way, too. A few examples will illustrate the process.

The living muskrat *(Ondatra zibethica)* is a rodent belonging to the vole-and-lemming group, the Microtinae. It is much the largest of the microtines, measuring 18–25 inches in length, of which 8–11 are on the tail. It has a dense brown fur, silvery white on the belly, and is highly valued commercially. The naked, scaly tail, which is compressed from side to side, is a characteristic feature. Muskrats live in marshes and at the edges of lakes and streams, where they feed on water plants and small animals, and build peculiar conical houses which may rear up to three feet above

The living muskrat, prime example of gradualistic evolution. It evolved from a Blancan ancestor in a sequence of species, grading imperceptibly from ancestor to descendant.

water. In this habitat they are common all over Canada and the United States except in the southeast, where their place is taken by the Florida water rat.

Thanks to patient research, especially by Holmes A. Semken, Jr., of Iowa City, the evolution of the muskrat is now known in surprising detail. The ancestor of *Ondatra* is a Blancan genus, *Pliopotamys.* Trends in the evolution of modern muskrats from early *Pliopotamys* include a steady increase in size and certain marked changes in the molar teeth. To understand their significance, it is necessary to take a look at the molar teeth of modern muskrats.

Like other rodents, the muskrats have a dental apparatus consisting of two distinct parts. In front are the incisors, ever-growing chisel-shaped teeth adapted to gnawing and biting. In the back of the jaws are the molars, which are used in mastication. As in other microtines, the molars are formed by a number of enamel-coated prisms, so that the wearing surface shows a series of triangles or loops, consisting of dentine bordered by enamel. In the angles between the prisms, there is third substance called cementum, which increases the triturating surface. A peculiarity is the presence of so-called dentine tracts, where the enamel is absent; they appear as V-shaped inlets from the base of the tooth. The effect is to interrupt the enamel lines, which enhances the chopping function.

The earliest muskrat known is *Pliopotamys minor* from the early Blancan of Hagerman, an animal the size of an ordinary vole. None of these trimmings are found in that species. There are no dentine tracts and no cementum. The late Blancan species *Pliopotamys meadensis* is somewhat larger and has incipient dentine tracts, but no cementum. Next comes a species, the Idaho muskrat, which is considered advanced enough to belong in the genus *Ondatra:* it has larger dentine tracts and the beginnings of cementum. The transition from *Pliopotamys* to *Ondatra* coincides with that from the Blancan to the Irvingtonian.

From *Ondatra idahoensis* of the early Irvingtonian, the evolution continued with size increase, development of larger dentine tracts, and more cementum. Three successive species are distinguished: *Ondatra annectens,* middle to late Irvingtonian; *Ondatra nebracensis,* latest Irvingtonian to early Rancholabrean; and the modern *Ondatra zibethicus,* late Rancholabrean to Recent. The history of the muskrats seems to be an excellent example of the gradualistic mode of evolution, for even within the

Many living species date back to the Irvingtonian, among them the masked shrew *(Sorex cinereus)*. Today a northern species, it ranged in the Pleistocene to Texas and northern Mexico.

successive species-stages, change may be observed from early to late forms, and it is very difficult to say where one species ends and the next one begins.

The fact that studies of this kind have been at all possible is in great measure due to Claude Hibbard, who pioneered the use of matrix-washing. With ordinary fossil-collecting methods, the minute teeth and bones of small animals like rodents, insectivores, and bats are very hard to spot. By washing and screening large amounts of matrix, such bones are often recovered in amazing numbers, and many a dump from earlier excavations has turned out to be a veritable goldmine in this respect. The invention has caused a revolution in vertebrate paleontology, and the previous focus on large animals has now shifted to the so-called micromammals. Humble though they may be in comparison with mammoths and giant sloths, these small creatures may yet teach us more about evolution and faunal history: they occur in great numbers and give great opportunities for detailed statistical analysis.

In the world of the micromammals, the Irvingtonian was an age of decisive modernization. Among the present-day voles of North America, the leading genus in terms of number of species is *Microtus.* Here belong such widespread species as the meadow vole *(M. pennsylvanicus)* of Canada and the northern United States; the long-tailed vole *(M. longicaudus)* of the West; the prairie vole *(M. ochrogaster)* of the Great Plains, and many others. The first *Microtus* voles appeared at the beginning of the Irvingtonian, 1.9 million years B.P., and more than twenty species are now known in the fossil state. The lemmings are a group of arctic voles, and they, too, date back to the Irvingtonian, with earliest representatives in the Cape Deceit fauna of Alaska. On the other hand, only a vanishing rearguard of the archaic Blancan voles survived in Irvingtonian times. And if we look at other rodents besides the voles, for instance the native American "rats," the story is much the same.

Not only the genera, but also the species of the Irvingtonian were, in many cases, identical with those still in existence. As we have seen, only a few Blancan species have survived to the present day. But in the Irvingtonian the number of modern species increased rapidly, so that at the end of the age the fauna is quite modern-looking, as far as the small mammals are concerned. This is in great contrast with the larger mammals, among which strange extinct forms were still predominant at that time. Indeed the later fates of the large and the small mammals have been very different: most of the former are now extinct, most of the latter are still thriving. This, again, is a topic which will be discussed in more detail later on.

The Coming of the Cold   49

IV
Full Blast

THE RANCHOLABREAN AGE

THE RANCHOLABREAN AGE may be defined as the time beginning with the first appearance of bison south of the land ice—probably during the Illinoian glaciation—and ending with the extinction of most of the large mammals (or "megafauna") at the end of the Wisconsinan glaciation. The precise date of the beginning remains to be determined. Obviously, the ice had to melt away, at least to the extent of opening a corridor between the Cordilleran ice in the west, and the Laurentide ice in the east, to permit immigration of bison. This could have happened during the Yarmouthian interglacial, or during a mild spell—an interstadial—of the Illinoian glaciation. The end of the Wisconsinan glaciation may be set at about 10,000 B.P.; that was the time of a decisive climatic amelioration, and the beginning of the present-day interglacial. As we shall see, the date 10,000 B.P. is also an approximate average date for the extinction of the large mammals.

The climatic history of the earlier Rancholabrean is still obscure. Analyses of deepsea cores suggest that there were several interglacials during the later Pleistocene, occurring about 100,000 years apart and lasting some 10–20,000 years each. In the classical scheme of North American glaciations and interglacials, the Sangamonian is the only interglacial falling in the Rancholabrean, but this is probably an oversimplification. In the absence of absolute dates, we cannot always be sure that what is called the Sangamonian in one place is the same as that in another: the radiocarbon dating method is only reliable for the Holocene and the later part of the Wisconsinan, and does not extend back to interglacial times.

The Wisconsinan was not a uniformly cold period. There were intervals when the climate was milder, although not as warm as in the interglacials, and these are called interstadials, while the cold phases are referred to as stadials. A similar sequence of stadials and interstadials occurred in Europe during the last glaciation. There is also evidence that earlier glaciations were interrupted by interstadials, but the dating methods are not precise enough, as yet, to enable us to map them in detail.

The geological history of the eastern Great Lakes and their surroundings suggests a threefold division of the Wisconsinan. It begins with the Early Wisconsinan stadial, a powerful ice advance producing a widespread mantle of glacial drift. Next came the St. Pierre interstadial, of which traces from the St. Lawrence Valley have been radiocarbon dated at about 66,000 B.P. The date is somewhat uncertain, being well beyond the limit for reliable radiocarbon measurements. The ice advance was resumed and interrupted by a second interstadial, the Port Talbot, perhaps at about 50,000 B.P. The Mid-Wisconsinan now continued with oscillating ice-fields and was terminated by still another interstadial, the Plum Point, about 30–25,000 B.P.

The Plum Point interstadial was terminated by the final or Main Wisconsinan ice advance. This stadial appears to have been colder than any of the previous ones with a culmination about and just after 20,000 B.P.

The Main Wisconsinan was the parting wallop, and a tremendous one. The amelioration to follow was, however, interrupted thrice by smaller advances of the ice occurring at about 14,000, 12,000, and 11,000 B.P. Each of these advances forms the termination of a short preceding interstadial called the Erie, Lake Arkana, and Two Creeks, respectively. One of the most famous early radiocarbon age determinations was made on the Two Creeks Forest Bed, a peat-bed with tree stumps near Lake Michigan in northern Wisconsin, overrun by the so-called Valders ice advance. Wood and peat from this site was dated by Willard F. Libby in his newly founded radiocarbon laboratory in Chicago in 1951, and the age obtained was 11,405 ± 350 years B.P. (The standard error given after the plus-minus sign represents a so-called confidence limit. It means that, in two cases out of three, the "true" radiocarbon age is between the limits thus given, in this case 11,055 and 11,755 B.P. It may sometimes be important to remember that in one case out of three, the error will be greater.)

This is hardly more than the bare bones of the climatic history of the last glaciation; interpretations vary, many dates need to be firmed up, and many details remain to be clarified. As far as the Main Wisconsinan goes, however, we are probably near the truth: an essentially similar story has been pieced together in Europe, for instance.

BERINGIA

Biogeographic evidence suggests that a land bridge in the Beringian area was in existence, intermittently at any rate, during the Tertiary period. From the great faunal interchange at the beginning of the Blancan we may conclude that the route was open at about 3.5 million years B.P. Soon afterward, however, it was flooded by the ocean, giving free access to marine animals between the Pacific and the Arctic seas: at that time we find exactly the same species of marine mollusks on Alaskan shores north and south of the Bering Strait.

In Pleistocene times, the major factor in the opening and closure of the Strait was the glacial-controlled rise and fall of the ocean, although movements in the earth's crust probably also played a part. At present, a fall in the sea level of 150 feet would suffice to open a narrow land connection between Chukotka Peninsula, on the Siberian side, and Alaska. An additional 15 feet would expose a second isthmus north of the Bering Strait, leaving a lake in the middle. If the sea level fell about 350 feet, which probably occurred during both the Illinoian and Wisconsinan glaciations, the entire Bering-Chukchi continental platform would be left high and dry. At that time, the land bridge was upwards of 600 miles wide at its narrowest part.

But should we not expect these arctic areas to be completely overrun by glacial ice, at the height of continental glaciation? Although it seems logical, such was not the case. Only very limited parts of Beringia were in fact covered by inland ice. The reason is probably very simple: the climate was too dry: there was not enough snowfall to feed a great glacier. In the same way, although northern Europe was covered by ice, no great ice-fields formed in Siberia, in spite of the cold.

In Beringia, patches of ice were present on the Chukotka Peninsula, and the Brooks Range in northern Alaska was heavily glaciated. Otherwise, the land bridge was ice-free. So the Siberian fauna could spread eastward into Alaska and even to the north-western part of the Yukon and Northwest Territories before being stopped by an impassable ice-barrier. Part of the barrier was formed by a lobe of the ice-field covering northern and eastern Canada, and part by a contiguous icefield based upon the Alaska Range and reaching well out onto the Aleutian island chain.

Climate in Beringia was harsh, yet there was sufficient vegetation to support a great number of land mammals, as we know from many find-spots in Alaska, the Yukon, and on the Siberian side. Many of the mammals were of Old World origin, but there were also American forms which had invaded the area in interglacial times and now remained marooned in Beringia. Here, then, mingled animals of different origin, some of them derived ultimately from very remote corners of the world—South America and Africa—while others had a long history in the Arctic. In fact, the Beringian area itself, with its intermingling of New World and Old World forms, and almost a subcontinent in itself, may have been an important evolutionary center.

A species of land animal migrating, say, from Siberia to North America would have to do so in two steps. It could enter Alaska only when the Bering Bridge was in existence, that is, during a glaciation. But it could not penetrate into America south of the ice before melting had proceeded sufficiently to create a corridor between the Laurentide and Cordilleran ice-sheets, i.e., at the beginning of an interglacial. In like manner, to reach Eurasia from North America, a species would have to enter Alaska during an interglacial, and "wait" for a glaciation to recreate the Bridge. It was a sequence of advances and burnt bridges.

All this may sound as if the migration were a purposeful undertaking, like setting out for Plymouth Rock, but that impression is wholly false. It is simply a question of "population spread," a much slower and more erratic process by which new areas at the margin of the range are invaded by vagrant individuals. In that way, a few rabbits introduced into Australia succeeded in populating much of the continent. Scandinavian zoogeographers have shown that the roe deer, beginning with a small local stock in southernmost Sweden around 1850, has now populated a great part of the country, spreading at an average rate of about 600 miles in a century. In the same manner Man entered Beringia, probably during the Wisconsinan, perhaps earlier.

We now have a tentative timetable for the Bering Bridge in the late Pleistocene. After flooding in the Sangamonian interglacial, about 100,000 B.P., sea level fell about 380–440 feet in the Early Wisconsinan. The milder conditions in the Mid-Wisconsinan caused the waters to rise again, and the Strait was recreated. Then, in the Main Wisconsinan, the sea level receded once more down to about 400 feet. About 18,000 B.P., the sea started to rise once more, and during the Erie interstadial, 13,000 B.P., the bridge was flooded briefly, only to be reestablished with the next advance of the ice. A second and longer flooding occurred in the Lake Arkana interstadial, but the bridge came into existence once more, and for the last time, about 10,000 B.P. Since then, there has been no land connection between Siberia and Alaska.

Ice-age Beringia, a great lowland area with waters encroaching and receding, was largely a herbaceous tundra. The spruce was almost gone: the timber line had been pushed far south and even the dwarf birches, willows, and alder bushes grew only sparsely in sheltered areas. Paul A. Colinvaux sums up the vegetation history which is spelled out in pollen diagrams from ancient lake-beds and marshes: the climate of the Wisconsinan was even colder than now. In spite of the continentality of the Beringian land mass, summers were shorter and colder than those of the present day. And the Arctic Ocean was covered by ice, as it is today: this had been doubted in a once-influential theory of the Ice Ages.

In wintertime, with temperatures down to −70° F, the ground was frozen. But summer would bring a thaw to the superficial layers, perhaps a few inches or feet, although lower down it remained frozen the year round: permafrost. On a hillside, the thawing, waterlogged topsoil would start to slide down, a phenomenon called solifluction or soil-creep. Remains of dead animals lying around would be picked up and enveloped by the sliding muck, finally to come to rest at the bottom of the valley, where the whole mass would gradually freeze again. Mostly the remains were bones, but occasionally the soft parts of a cadaver, or even a whole animal, might be caught up and imbedded before being dealt with by carrion-eaters. The exposure to the cold, dry climate would already have desiccated the corpse into a mummy; now, with solifluction muck piled on top, it came within the range of the permafrost. And so, as nuggets in the goldminer's harvest, occasional dry-frozen bodies would turn up among the bones.

Talk of goldminers in this context is not idle, for most of the fossils, for instance from the Fairbanks area in Alaska, and Dawson City in the Yukon, have been found by them. The stream-beds in the valleys, beneath the solifluction deposits, are where the gold is found. To get at them, the overlying muck-beds are hosed off with powerful jets of water, and in that process thousands of fossil remains are unearthed. Cooperation between miners and fossil collectors has resulted in many remarkable discoveries.

Of course the most conspicuous bones were those of great creatures like mammoths and bison, and it is no wonder that tales about "giant men" found in Alaska were rife in earlier times. They still tend to crop up. I recently happened to read one in which it was stated at the end that "scientists" refuse to believe and publicize such stories because they "do not fit in with the scientist's preconceived ideas."

Other myths have been inspired by the mystery of the frozen carcasses. There is, for instance, a story about a mammoth-steak banquet in connection with the excavation of the famous specimen from Beresovka River in Siberia, now on exhibition in the Zoological Museum in Leningrad. As far as I have been able to ascertain, the facts seem to be that one of the members of the expedition made a heroic attempt to eat a piece, heavily seasoned, but was unable to do so.

The most influential myth at present dates from 1960 and proceeds on the assumption that mammoth bodies were frozen with extreme rapidity in an extraordinary climatic catastrophe—or else they would show signs of decay. But it has been well known—at any rate since the Russian zoologist Otto Herz published his enthralling account of the Beresovka Expedition (in 1902)—that the carcasses *were* partly decayed before freezing; and so the whole rigmarole collapses. Recently, Michael R. Zimmermann and Richard H. Tedford studied the tissues of some Alaskan specimens—among them a hare, a lynx, and a baby mammoth (dated at 21,300 B.P.)—and found much of the original matter replaced by masses of bacteria, especially in the inner parts where the cold took longer to penetrate. They conclude that after death the remains were partly decomposed, and often dismembered by scavengers, before being entombed by the normal depositional processes. Of course, the idea of a single catastrophe is negated by the varying ages of the finds—frozen Siberian mammoths, for instance, varied in age from 11,400 B.P. to 44,000 B.P. and beyond.

BERINGIAN LIFE

Ice-age Beringia may seem to us an utterly inhospitable world; yet we know that it supported a great range of fauna, and even human beings lived there.

Most of the animals whose remains are now found in the frozen muck are ones that we immediately associate with an arctic or boreal environment. Even a partial list will show this: mammoths, muskoxen, caribou, tundra hares, lemmings, arctic foxes, wolves, wolverines, grizzly bears, Canada lynxes, Dall sheep. But there were also creatures which were not primarily adapted to a cold climate; they entered in interglacial times and managed to survive. The most amazing one is perhaps the Jefferson sloth, *Megalonyx jeffersonii.* It is a far cry from the tropical rain forest of the living tree sloths to the Alaskan tundra of that ground-living giant.

A special group is formed by those Old World species which managed to reach the American part of Beringia but were unable to conclude their conquest of the new continent. One of them is the yak, a great highland ox which is still found in the wild state in Tibet, and is also known as a domestic animal. Domestic yaks are medium-sized cattle, variously colored. The great wild form, reaching over six feet at the shoulder, is jet black with a white dorsal stripe, and has long, curved horns. It is very well adapted to its dry, cold, wind-blown homeland, and so is not an unlikely invader of Beringia. Its mode of life may give an idea of how the Beringian plant-eaters managed to survive. In the summer, the yaks are found in the mountains near the glacier margins where the meltwater supports a comparatively lush vegetation. Their winter pasture is on the plains, where they manage to uncover the sparse hay by scraping away the snow. From year to year they follow the same paths of migration, and they are trampled into veritable roads.

The saiga antelope is another Asiatic animal whose advance into North America was stopped by the ice. Surviving now in an area extending from the Russian steppes in the west to Mongolia in the east, it had a much wider distribution in the Pleistocene, ranging from western Europe to Alaska and the Northwest Territories. Related to sheep and goats, it is easily recognized by its peculiarly swollen muzzle and lyre-shaped horns.

Oddly enough, one of the most characteristic arctic mammals of the Old World never seems to have penetrated into Beringia. This is the woolly rhinoceros, *Coelodonta antiquitatis,* a comparatively small, heavily furred rhino with tandem horns on its nose. Its eastern boundary seems to have been at about 165° eastern latitude: it has not been found beyond the Kolyma river valley. Why this should be so is difficult to understand, and there is always the possibility that future finds may add to its known range.

In Eurasia, the woolly mammoth and the woolly rhino have almost identical ranges, but only the former is known to have invaded North America. It was present in the

Fairbanks area from the early Rancholabrean on, and evidently spread into other parts of arctic North America in the Sangamonian. In Wisconsinan times it retreated south before the advancing ice sheets, and is found in the area close to the ice margin in southern Canada and the northern United States—the so-called periglacial area. Seasonal migrations probably occurred, so that it penetrated farther south in wintertime.

The scientific name of the woolly mammoth is *Mammuthus primigenius,* the "first-born mammoth." Actually it was the last of its line: the Eurasian mammoth line, starting, like the American one, with *Mammuthus meridionalis.* The two lines run parallel in some respects. As in American mammoths, the cheek teeth became very complex, consisting of thin transverse plates—the very long last molars may consist of up to 27 such plates, closely appressed, which is more than even in the most advanced American mammoths. As in all elephants, only one molar at a time is in function in each jaw half; the next molar comes into play from the back of the jaw as the previous one is worn out. Each molar is larger than its predecessor and elephants tend to grow in size, too, almost throughout life.

Contrary to its popular image, the woolly mammoth was not especially large, as elephants go. Its Eurasian predecessors, the early *M. meridionalis* and a transitional species, *M. armeniacus,* were much larger and the latter may have been the biggest elephant ever in existence. But the late Pleistocene *M. primigenius* was rarely more than ten feet tall.

Its appearance must have been awesome. It is known to us, not only from the frozen cadavers but also from eyewitness portraits, the paintings and engravings of Stone Age man. It had a very high, peaked head and a sloping back with a pronounced hump in the shoulder region—not a fat hump as in camels, but a structure formed by the curvature of the backbone. The tusks were immense and strongly curved. The ears were quite small, to reduce heat-loss (the African elephant uses its large ears to get rid of surplus heat), and the trunk, probably for the same reason, was comparatively short. It had two movable fingers at the tip, a precision-grip apparatus which may have compensated for the shortness of the trunk. Insulation against the cold was provided by a subcutaneous fat layer about three and a half inches thick, and by long hairs and a thick, woolly undercoat. As preserved, the hairs vary from black to yellowish; but Kenneth P. Oakley at the British Museum thinks that the original color may have been uniformly black, the tints now seen resulting from degradation of the pigment.

Yaks and caribou have to scrape away the snow with their feet to reach their fodder. The mammoth could, and apparently did, use its tusks. Many of them show wear on their under surface, and we may imagine the mammoth tossing the snow to one side with powerful heaves. The problem was solved in yet another way by the woolly rhino, whose long, almost bladelike horn, compressed from side to side, forms an efficient snow shovel; it, too, shows the telltale wear facet, all along the front edge of the horn. The struggle for existence in a bleak world is brought to us vividly by these remains of animals that vanished from the earth thousands of years ago.

Woolly mammoths

The woolly mammoth is the embodiment of the Ice Age. Long will it live in our imagination, a black, top-heavy shape looming up in the swirling snow, great tusks gleaming: to our forefathers, perhaps, a demigod and a living storehouse of meat at the same time.

While the mammoth is gone, two other plant-eaters of the High Arctic survive today: the muskox and the caribou. The muskox is the sole survivor of a once-varied group of species, some of which we have already met in Irvingtonian times. The true muskox, *Ovibos moschatus*, dates back to the early Rancholabrean in Alaska. It managed to penetrate south of the ice well before Wisconsinan times, its range extending from England in the west to Illinois and Nebraska in America. It survived within most of this range up to the end of the Ice Age. It is possible that the muskox was one of the very few mammals which managed to survive in the arctic of Canada, north of the ice sheet, during Wisconsinan times.

The modern muskox is probably more thoroughly adapted to arctic conditions than any other of its tribe. It is a moderate-sized ox, reaching a shoulder height of about four and a half feet, short-legged and stocky of build, with flattened, down-sweeping horns. Its fur gives it an extremely efficient insulation against the cold, for in addition to the long guard hairs, which hang to the ground, it consists of a very fine, soft undercoat, almost impermeable to moisture. Its gregarious habits are important, too: in a storm the oxen flock together and form a protecting wall for the calves. The same method is used in defence against predators, but this, unfortunately, does not help when man is the attacker. In fact it may be disastrous, for the muskoxen refuse to budge even though their mates are killed right and left. This misguided instinct may have contributed to the extinction of the species over much of its range after the Ice Age. The last Eurasian muskoxen probably died out some 3,000 years ago, and in North America their numbers have diminished greatly in historical times.

At least two other kinds of muskoxen inhabited North America in the Rancholabrean. The woodland muskox, *Symbos cavifrons*, which appeared in the late Irvingtonian and survived to the end of the Wisconsinan, was taller and more slenderly built than *Ovibos*. It was evidently not an exclusively arctic mammal, for it ranged from Alaska into California and Texas; the total number of finds is about ninety.

The second type of muskox, *Bootherium bombifrons*, is much rarer. The original specimen was collected at Big Bone Lick in Kentucky by a party sent out by Thomas Jefferson; later on, some remains were found in a cave near Frankstown, Pennsylvania, and recently others have been found in Idaho. This little-known animal was smaller than the living muskox.

The caribou or reindeer, *Rangifer tarandus*, may have originated in Beringia or in the mountains of northeastern Asia. It was present in Beringia well back in the Irvingtonian, for it occurs in the Cape Deceit fauna by Kotzebue Sound. In Wisconsinan times this arctic grazer ranged as far south as Tennessee; today, the extreme southern populations of caribou are found in northern Idaho, the northern Great Lakes region, and in Maine.

The most commonly found animal in Alaska's frozen tundra is the bison, of which not only bones but also mummies have been unearthed. The fossil bison of Alaska are highly varied and have been referred to several species, but it seems possible to interpret most of them as representatives of a single evolving species, *Bison priscus*. It ranged over most of Eurasia too, and was portrayed by Paleolithic man in Europe; the most famous paintings are in Altamira Cave in northern Spain. The combined evidence from cave art and frozen carcasses gives a good idea of what the animal looked like.

There are two living species of the genus *Bison*. The American buffalo, *Bison bison*, once close to extinction, is now protected and its future appears to be safe. The European wisent, *Bison bonasus*, survives in a semiwild state in a protected forest area shared by Poland and the Soviet Union; additional populations are in zoos.

Bison priscus, larger than the living species, looked rather different. Living bison are short-horned; *Bison priscus* had long, mostly gently curved horns. In cave art, the horns are depicted as deflected backward at the tips. This characteristic is not seen in the bony horn-cores, and so it has been thought that the Paleolithic cave painters used their artist's licence to embellish the picture. But the mummies, in which the outer sheath of the horn is preserved, show that the tip was in fact curved in precisely the manner shown in the paintings. The horns are particularly long in the earlier forms but tend to become shorter, and also more curved, in later bison.

Another characteristic in which the fossil bison differs from the living is the mane. *Bison priscus* had a very conspicuous, black dorsal mane on the back of the neck and behind the shoulders, divided by the shoulder hump. There was also a ventral mane extending from the throat to the belly. Together, these body ornaments make a very striking pattern, visible and identifiable at a great distance. For a herd animal living in the open plains this could be a useful characteristic: they would be able to recognize their companions from afar.

The Rancholabrean bison of North America south of the ice are descendants of the Alaskan *Bison priscus*. There must have been numerous migrations and a repeated mixing of stocks. And yet the Alaskan-Eurasian species seems to retain its identity, almost up to the end of the Wisconsinan. The last occurrence of the big-horned Beringian form, on Old Crow River in the Yukon, is radiocarbon dated at 11,910 ± 180 B.P.

LION TRIUMPHANT

When we speak today of American lions, we usually think of the animal variously called mountain lion, cougar, puma, or (in Florida) panther. There is some resemblance. Like the lion, the puma is self-colored; the fur varies from grayish to a tawny tint not seen in true lions. The cubs are spotted, in the puma as in the lion. But the puma is much smaller than a lion, and is not now regarded as a member of the genus *Panthera*, which comprises most of the great living cats, including lions and (true) panthers.

The lion, today, is an almost exclusively African animal. There is a small remnant of the Asiatic lion, *Panthera leo persica*, in the Gir Forest of Kathiawar, India, where it is protected. In Iran, protection came too late to save the species; there it was last heard of in 1942. The famous Barbary lion of North Africa, *Panthera leo leo*, has been extinct since the turn of the century. Various other subspecies are still in existence, locally in large numbers, in other parts of Africa.

In antiquity, the lion ranged north into Greece and Asia Minor, as we know from historical records, pictorial representations, and bone finds. There is many a reference to the lion in Homer, and its bones have been recovered from Troy. In Mesopotamia, lion representations are common, for instance in Assyrian art.

The great cave lion of Ice Age Europe, *Panthera leo spelaea*, was first described in 1810. Its bones have been found at many sites, especially in caves, and it ranged well into England. There has been much speculation as to whether it was in fact a lion, a tiger, or something else; one author even maintained that giant tigers and lions alternated in Europe. But the consensus among serious students now is that the European form was indeed a lion, and most regard it as a subspecies of *Panthera leo:* a race acclimatized to the rigorous conditions of the Ice Age.

That a great cat can exist in a northern climate is well proved by the present-day Siberian tiger, largest of living cats.

In 1853, Joseph Leidy, still early in his career as a pioneer of vertebrate paleontology in the United States, described the gigantic cat *Panthera atrox ("Felis" atrox)*, based on material from Natchez, Mississippi. At once the debate on the relationships of this animal was initiated, and it has gone on to the present day. But here, too, a consensus has been reached. Experts agree that it is a lion, closely related to the European cave lion; possibly a species of its own, but more probably just a subspecies, *Panthera leo atrox.* It has now been found at about forty sites, ranging from Alaska to Mexico, from California to Florida.

Much of the story of the lion can now be pieced together. It started in Africa about two million years ago, as suggested by finds from caves in the Transvaal and from Olduvai Gorge in Tanzania. About 700,000 B.P., the species spread into Europe, starting upon a career that was to be unique in the history of wild mammals. These early European lions are outsize, even for cave lions, so it seems the species attained its greatest size at this time. On the other hand, these lions were comparatively small-brained. In later European and African lions, brain size increased, and Helmut Hemmer thinks this may be connected with the appearance of social habits. Most cats are solitary, but lions form "prides" which hunt in tactical cooperation. This is reflected in brain size: lions have larger brains than solitary great cats.

Soon, the lion also spread into eastern Asia. There is at least one find from the famous Peking Man site of Choukoutien, where otherwise the predominant big cat is the tiger. This find may be about 400,000 years old. Then, various later Pleistocene finds record the presence of lions over most of Siberia. In 1971, the Soviet paleontologist N.K. Vereshchagin listed a total of fourteen fossil lion sites east of the river Lena. Of these, two are on the New Siberian Islands right out in the Arctic Sea, one near Vladivostok on the Pacific coast, and no less than six in the Kolyma River Valley at the border of Beringia itself. And the Beringian record continues on the American side in Alaska and the Yukon.

Already at this stage, the lion had conquered a range which is remarkable for a wild animal. But it did not end there. Once more the icefields of North America wasted away, and lions advanced to the south, into the province of Alberta and into the United States. By Sangamonian times it had reached the Valley of Mexico. In the Wisconsinan it seems to have been present all over the continent south of the ice, except perhaps some densely wooded areas like the Florida Peninsula.

And still the advance went on. Lions pushed through Mexico, crossed the narrowing Central American land bridge and spread further south along the Pacific coast. We know they reached Peru, for lions were found in the tar seeps of Talara in the northwestern corner of that country. This is the high-water mark of leonine conquest.

At the height of its triumph, then, the lion was present in five continents: Africa, Europe, Asia, North America, and South America. For a wild animal, this story is

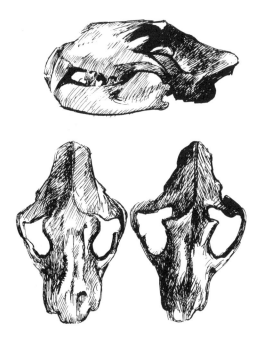

Lion skulls. Top and left, American lion, *Panthera leo atrox;* right, early European lion, *P. leo fossilis.* Note the broader skull and larger braincase of the American form.

unparallelled in the history of the earth. No other species of land mammal has ever conquered such an area before the coming of Man with his parasites and domestic animals. Well may the lion be styled the King of Beasts.

To perform this feat, lions had to adapt to almost every climatic regime imaginable, from the dry cold of the Arctic to the humid heat of the tropics; to make a living in rain forests, on sun-scorched steppes, in temperate forests, in the mountains, on the northern tundra. The versatility of the species boggles the mind. Explanations are not easy to find.

The lion, of course, is a carnivore, and as a rule, carnivore species are more wide-ranging than herbivores. Plant-eaters usually specialize in certain kinds of vegetation: the bison keeps to the plains, the moose to the forest. The lion can prey on either. In fact the most wide-ranging living land mammals are Carnivora: the red fox, the wolf, and the brown (and grizzly) bear.

Then there is the evidence, noted above in the case of the Siberian tiger, that some big cats may have a very wide climatic tolerance. The present-day puma ranges from southern Canada into Patagonia. During the Wisconsinan glaciation, jaguars lived as far north as Nevada, Kansas, and Missouri, and at the same time there were leopards in central Europe.

But there may be more to it. As Helmut Hemmer has pointed out, the lion is an unusually large-brained cat, and this may be connected with its social habits. The lions of the North were larger-brained than those living now, and the brainiest of all lions was the American one. Hemmer calls it a supercarnivore, not only in terms of size and strength, but also in terms of intelligence, as suggested by its superior brain

A fragment of a reindeer shoulder-blade from an ice age cave in France shows the tufted tail of a lion.

size. Perhaps it is this combination of advanced characteristics which lies behind the achievements of the Pleistocene lion.

What did the American lion look like? It certainly averaged larger than any living subspecies. The head-and-body length, not counting the tail, has been estimated at between five and a half feet (c. 165 cm) in small females, and eight and a half feet (c. 255 cm) in large males.

Unfortunately there are no contemporary pictures of *Panthera leo atrox*. On the other hand, various European engravings and figurines portray large cats, and some of these are clearly lions. An interesting find is an engraving showing the entire tail, with the characteristic lionlike tuft. It is on a piece of reindeer shoulder-blade from France, but sadly enough the main part of the picture has been lost. At any rate it shows that the cave lion had a tufted tail, which had been doubted because the tuft is absent in some other pictures of big cats.

Some of the lion pictures show markings on the body which have been interpreted as stripes or spots, and it is possible that the spots present in cubs may have persisted in adult animals, as they occasionally do in modern lions.

The male lion, as everyone knows, is distinguished by its mane. Hemmer thinks that a very dark type of mane, with marked extensions along the back, behind the shoulders and especially along the belly, represents an early condition. Contrasting sharply against the light color of the ordinary pelage, it would give the animal a strongly patterned exterior serving as an identification marker in open country with long visibility. This type of mane persisted in the Cape lion of South Africa and in the lions of the North African deserts, and much the same type is also seen in the ancient lion pictures of Asia Minor and the Levant.

The mane is less developed in the modern lions of eastern and western Africa; there is usually no mane behind the forelegs and along the belly. In this, judging from pictures, they resemble the lions of classical Greece; there is no real belly mane although the hair on the belly is long. Probably the same was true for the cave lions, in which the mane appears to have been even less developed, and perhaps had a lighter, less contrasting color. But whether this was true for the American lion, too, is anybody's guess. Perhaps, one day, the discovery of a frozen lion will give the answer, as in the case of the bison. If so, my bet is that it will happen in Alaska, where the immense harvest of lion bones shows it to have been very common.

THE SEAS OF BERINGIA

In the 1950s, Maurice Ewing and W. L. Donn, of the Lamont Observatory of Columbia University, presented a most ingenious theory of the Ice Age. In a nutshell, it visualized a cyclic feedback between the Arctic Sea and the surrounding land masses. In the early stage of a glaciation, the sea was ice-free; evaporation brought increased precipitation to the surrounding lands, which were glaciated. Increasing cold and reduced water circulation then led to the sea freezing over; precipitation was reduced and glaciers dwindled. Interglacial conditions returned and the sea-ice melted (the current interglacial has not yet reached this stage). So the stage is set for a new cycle.

The theory, however, has not fared well; paleoclimatologists find that the Arctic Sea probably has been ice-bound constantly for at least 150,000 years. But for how much longer? Knowledge of the marine fauna of the Beringian seas is of particular interest in this connection.

Unfortunately, although we have so much evidence of the land fauna, relatively few mementoes of the teeming life off Beringia's shores have come down to us. Only during the warm interglacial phases did the sea encroach upon the present-day land areas, and left deposits in which remains of its life are found.

The microfossils and the remains of marine shellfish, of course, tell much of the story. But the mammals living in the sea may also be important. One of them is the sea otter (Enhydra lutris), a large, shellfish-eating mustelid which in historical times ranged from the Aleutian Islands to Baja California. Severe hunting brought this valuable fur-bearer to the brink of extinction; by now, however, last-minute protection has already saved the species and the sea otter is now increasing.

The sea otter seeks his food (mainly sea urchins in Alaskan waters) on the bottom of the sea, down to about one hundred feet. Returning to the surface with an armful of food, he will then lie on his back, spreading his meal over the dinner-table furnished by his own belly, and methodically crack one urchin after the other. Harder shellfish are split open with a specially procured stone, probably the nearest approach to cutlery among non-human mammals. Occasionally, the otter has to fend off hungry gulls, who have no table manners. He does it by splashing water in their faces.

The life-zone of this most inoffensive and endearing animal is restricted by sea ice, so we may be sure that its presence indicates open water. Now, sea otters have not been found in the Rancholabrean of Beringia, although its remains have been identified in deposits further to the south. But there is one intriguing find from Point Barrow, the very northern tip of Alaska, right in the Arctic Ocean.

Charles Repenning, a U.S. Geological Survey paleontologist, and an expert on the history of Beringian mammals and of marine mammals in general, thinks this find may be very old; it could date from the Yarmouthian Interglacial (late Irvingtonian). Whatever its date, ice conditions must have been markedly less severe than now. And so the Point Barrow sea otter may be taken as one of the items of evidence suggesting that the Ice Age reached a climax only in its later part, in the Rancholabrean, and that the Arctic Ocean was at last partly ice-free in the Irvingtonian.

Seals of many kinds have been taken in Rancholabrean marine deposits in Alaska—fur seals, sea lions, harbor seals, ringed seals, and bearded seals. Walrus have been found at two or three sites, the oldest being at the Kokalik River near Point Lay, and probably of early Rancholabrean date. From here it ranged to California; but walrus fossils are much more common on the Atlantic seaboard. Toward the end of the Wisconsinan, and in early Postglacial times, walrus inhabited the Champlain Sea—a precursor of today's Great Lakes—and two bones have even found their way into eastern Michigan, apparently carried there by man.

Perhaps the most remarkable animal of the arctic waters, however, was Steller's sea cow *(Hydrodamalis gigas)*, which survived all the vicissitudes of the Pleistocene and Postglacial, only to be exterminated by eighteenth-century man. It was discovered by Vitus Bering's famous 1741 expedition and studied by the ship's naturalist, G. W. Steller. In less than thirty years it was gone, and its remains now belong to the greatest rarities in natural history collections.

The sea cow, like the Florida manatee, belongs to the mammalian order Sirenia. Like the whales, sirenians are so far adapted to life in the water that they cannot survive out of it. Again like the whales, they have a streamlined body; forelimbs transformed into flippers; a horizontal fluke on the tail; and no hindlimbs. Yet their closest relationship seems to lie with proboscideans rather than whales, as suggested by various anatomical details. The manatees even have a back-to-front replacement of the teeth somewhat like that of elephants. (The sea cow, on the other hand, was nearly toothless.) The fanciful name of the order, Sirenia, alludes to the bewitching sea-maids of the Odyssey. In fact, the female has two breasts in about the human position, and

Steller's sea cow, the last of the great Pleistocene mammals to become extinct. (After an early sketch by Steller)

sometimes holds its nursing baby in place with its flippers. Apart from this, I am sad to say, the resemblance is remote . . . to say the least.

Steller's sea cow was the biggest of the Ice Age sirenians, and the only Arctic member of its tribe. Attaining a length of upwards of 25 feet, this giant may have weighed as much as ten tons. Beneath the peculiarly barklike skin there was a thick layer of blubber, much coveted by whalers. In combination with the animals' habit of browsing in herds on the water vegetation at shelving, sandy shores, this proved to be their undoing, for they were easily hunted.

The eighteenth-century herds of sea cows around the Commander Islands, between the Aleutians and Kamchatka, was only a remnant of a much greater Pleistocene population. As late as 19,000 B.P. the species ranged to Monterey Bay, California, and its history goes back to an even more widespread Miocene genus, *Metaxytherium*. Possibly it would have become extinct even without human interference; on the other hand, modern conservation measures might have saved it. Less fortunate than the sea otter, Steller's sea cow became the last of the great Pleistocene mammals to die out.

THE PLATES

Before the Indians: A Pictorial History

In this portfolio of reconstructions, spanning more than three million years of America's past, Margaret Newman and Hubert Pepper unfold a broad panorama of changing scenes and actors. Starting with the pre-Ice-Age world of the Blancan age, they then proceed to the Irvingtonian, with its rejuvenation of the stage resulting from the first continental glaciations and a burst of intercontinental migrations. The remaining and major part of the paintings recreate America in the Rancholabrean age, from Alaska in the full blast of the cold to the southern refugia of Florida and California, where persisting warm climates permitted the mingling of subtropical and temperate-climate forms.

Most reconstructions of this type tend to focus on scenes of violence and drama, depicting the most spectacular of those beings of bygone times—"Nature red in fang and claw." That element is here, too, but the artists have also recaptured the everyday life, and the minor actors as well as the stars. Here you will find not only the mammoths and giant sloths, sabertooth cats and hunting bands of early men, but also the prairie dogs and owls, the snakes and tortoises, and the landscapes with flowers and cacti, swamp cypresses and palmettos, all based on meticulous research and set forth with thorough competence and artistic skill.

To explain how it was done I can do no better than quote Margaret Newman, who describes her approach as follows:

Recreating the past is an exacting form of art. At no time can one allow oneself "artistic licence." The artist should be prepared to question every detail of his proposed reconstruction. Beginning with the landscape, he must establish the nature of the physical conditions prevailing in the locality and geological age to be reconstructed, then obtain photographs of a region in which such conditions exist today. If it is possible to visit the area, so much the better. It is not sufficient merely to consult the work of another artist—he may be equally uninformed.

The work may now proceed in the following sequence:

First, the ancient topography and climate should be established. Both may be utterly different from conditions in the area today. The composition of the rock in which the fossils are found will give good clues. For instance, shales and certain types of sandstones consist of clays and sands which were deposited by slowflowing rivers or in lakes, indicating low-lying country with a moderate to high rainfall. Sandstones consisting of wind-blown sands or conglomerates of rocks, stones, and sands are often representative of desert conditions, while glaciation of an area is indicated by the presence of ice-scratched rocks, erratic boulders, moraines, and other diagnostic glacial features. The nature of igneous rocks and volcanic ash deposits indicate the type and extent of volcanic activity at the time of their formation.

Next comes the flora. The vegetation of an area is largely influenced by its topography, climate, and soil type, all of which will already have been established. If it is necessary to show the plant species in detail rather than as a distant green blur, a study of the fossil flora of the locality is required. Unfortunately conditions favoring fossilization of animal remains are quite often unfavorable to the preservation of plants. As a locality is usually chosen for display on the basis of its paleofauna rather than its paleoflora, there may be no plant remains known from that particular site. In this case it is necessary to refer to the nearest locality of the same age, topography, and climate from which plants have been recovered and assume that a similar flora may be used.

Turning now to the fauna, the mammals, in the present case, form the main subject of the reconstructions. Generally speaking the choice of animals to be depicted is limited by the number of actual finds or to the predominance of a few particular species. Sometimes, however, the fauna assemblage is so large and diverse that it is difficult to decide which species are of chief importance. I have been fortunate in working with a paleontologist who has a practical approach to such a situation, but many artists encounter considerable difficulties with a scientific supervisor. In his anxiety to portray to the public the wonderful diversity and richness of the fauna he may well order as many as twenty different species to be included in one scene. The artist then finds himself in a dilemma. The animals must be given something to do if the reconstruction is to retain a certain degree of animation and interest. The resulting scene is usually one of unbelievable savagery in which every animal is either chasing or being chased, devouring or being devoured. A visit to a game reserve should be sufficient to convince the artist that such a gathering of species and such scenes of mass carnage rarely occur.

It should be possible to introduce as many as six species into one picture with the use of a little imagination, but no more if the reconstruction is to remain credible to the viewer.

The reconstruction of an extinct animal depends largely on the amount of fossil material available. Nobody should be expected to attempt such a task on the basis of a single bone. If insufficient material is available from a chosen locality, it may well be that the same species is better represented in other collections and it is advisable to check all the literature on the subject and to contact the recognised authority on the particular species in order to make sure that all the available information has been collected.

Assuming that enough skeletal remains have been found, a muscle reconstruction can be made by careful study of the bones. Knobs, ridges, and roughened areas on the bone surface usually indicate the points at which muscles were once attached. Reference to anatomical drawings and dissections of closely related living forms may be used to build up the complete musculature. The skin depends much on the type of animal, its way of life, and its habitat. Once again the study of living related forms or those living in a similar habitat may be of use. Camouflage coloring such as spots or stripes may add interest and authenticity provided that it is appropriate to the type of environment in which the animal lived. Modern elephant and rhino tend to adopt the color of the ground in the area in which they live due to the mud-wallowing and dust-bathing habits. They can therefore be portrayed accordingly in their fossil forms.

Extras, finally, include details which, although not necessary to the overall scene, do help to give authenticity and familiarity to the whole. Insects and other invertebrates may be added to an otherwise vertebrate fauna with effect, provided that their remains have been recorded from the locality or from one of similar age and paleoenvironment. Footprints, mud-cracks, and fallen leaves are useful as a "foreground interest," while a half-buried skeleton performs a double function by suggesting the method of fossilization and by introducing into the scene another species for which an occupation does not have to be found.

The artist has to adapt to many fields of work, all of which demand accuracy as well as creativity. He cannot be dismissed with that time-honored derogation "clever with his hands." He must be prepared to acquire at least an elementary knowledge of geology, paleontology, archeology, zoology, botany, and ecology and finally must be able to present his work in a way that his experience tells him will be both pleasing, informative, and above all credible to his public.

Plate 1

The time before the first great glaciations in America, corresponding to the later half of the Pliocene epoch in earth history, is called the Blancan age. It commenced some 3.5 million years ago and lasted till about 1.8 million years B.P. Its beginning is marked by the appearance of various new kinds of mammals—for instance, the deer genus *Odocoileus* (to which belong the present-day white-tailed and mule deer), the *Geomys* genus of pocket gophers, and the cotton rats or *Sigmodon*. Above all, they include the first modern-type, one-toed horses: the American zebra *Equus simplicidens*, a close relative of the living African species, Grévy's zebra (Plate 1). It was the first Blancan guide fossil to be regarded as such. Other characteristic Blancan species include a cheetah *(Miracinonyx studeri)* and a bone-eating dog *(Borophagus diversidens)*, also shown in the painting. The presence of cheetah, a fast-running plains form, indicates an open landscape. In this scene, the zebra is more concerned about the advancing *Borophagus* dog than about the cheetah, which is obviously not hunting. The stage represents Cita Canyon, a tributary of the Palo Duro Canyon, Texas.

Plate 2

In this scene, based on finds from Mt. Blanco, Texas, a hunting hyena *(Chasmaporthetes ossifragus)* is closing in on its prey, a three-toed gazelle horse *(Nannippus phlegon)* on the Blancan plain. Finds of a hyena in America were first met with incredulity, for this is a typical Old World family of carnivores. The species, which has close relatives in the Old World, is now seen as one member of the great trans-Beringian immigration at the beginning of the Blancan. Its slender limbs indicate a fast-running form which may have hunted in the manner of cheetahs or today's African hunting dogs, but it was not specialized for smashing bones like the living hyenas. That niche was filled by *Borophagus* (Plate 1). The gazelle-horse *Nannippus*, a slender creature about 3 feet high, was a surviving three-toed horse; it became extinct toward the end of the Blancan. Again, the presence of these fast-running species indicates open country.

Plate 3 *(see next page)*

The Irvingtonian age started with a new phase of intense intermigration between Asia and North America, across the temporarily emergent Beringia. Among the migrants may be noted the southern mammoth *(Mammuthus meridionalis)*, a species which had evolved in Eurasia. After entering the New World, it gave rise to a long line of American mammoths.

In this scene, based on the early Irvingtonian (c.1.5 million years ago) site of Rock Creek, Briscoe County, Texas, a southern mammoth with its yearling calf is shown together with an immigrant from South America, the heavily armored *Glyptotherium arizonae*, on a river bank. The glyptothere, a distant relative of the armadillos, was a huge animal, measuring some 10 feet in total length and weighing about a ton; only the carapace and the banded tail are visible. On the other bank, another mammoth and a group of horses *(Equus scotti)* may be seen. A swimming water rat *(Neotoma taylori*, an extinct species) is betrayed by its wake. Cornflowers are in bloom.

Plate 4

Prior to the Irvingtonian age, no member of the great bovid family (cattle, bison, sheep, goats, true antelopes, and so on) are known in America: this group of mammals evolved in the Old World. The first bovids reached America at the beginning of the Irvingtonian, among them the shrub-ox *(Euceratherium collinum)* shown here. Now thought to be related to the muskox, it differed in the higher position of its horns. It probably inhabited the lower hills, as suggested in this scene showing two competing males threatening each other. The place is near Irvington, California, in late Irvingtonian times, around 800,000 years ago. Canada geese have also been found at this site.

M. LAMBERT NEWMAN

Plate 5

Scene near Cumberland Cave, Maryland, in late Irvingtonian times, shows a flying squirrel *(Glaucomys volans)* making its escape from a marten (upper left). This species of squirrel, which appeared in the late Irvingtonian, survives to the present day. The marten shown is an extinct form *(Martes diluviana)* related to the modern fisher. Clinging to a branch is a small, extinct porcupine *(Coendou cumberlandicus);* living coendous are found in tropical forests from Mexico to Brazil. The bats are eastern big-eared bats, *Plecotus alleganiensis;* again, this is an extinct species, but closely related to the living *Plecotus rafinesquii.* Remains of 49 species of mammals were found in the cave fill. The original entrance to the cave was on top, and it acted as a trap with a 100-foot drop. The tree is a shagbark hickory.

Plate 6

The Irvingtonian was the time of the two first continental glaciations in America —the Nebraskan, beginning about 1.5 million years ago, and the Kansan, beginning 900,000 years B.P. Deposits near Medicine Hat, Alberta, record the life of the intervening interglacial. By this time, the southern mammoth stock (see Plate 3) had already given rise to another species, the Columbian mammoth *(Mammuthus columbi)* shown here. The mammoth, in the act of beating a shrub against its foreleg to shake off the earth, is startled by a prairie dog *(Cynomys ludovicianus)* popping out of its burrow to stand sentinel. This species of prairie dog is still extant.

M. LAMBERT HEWMAN

Plate 7

Plates 7 and 8 show a landscape during and after a glaciation to explain some of the geological features resulting from the action of glacial ice. In Plate 7, a piedmont glacier has developed, originating from three cirques (also called corries), bowl-shaped recesses excavated in the mountainside by glacial action. The streams of ice are bordered by lateral moraines, consisting of eroded rock material transported with the ice; where the two streams unite, their lateral moraines combine to form a medial moraine. Crevasses open and close as the ice moves; a very marked crevasse, termed a bergschrund, separates the moving ice from the mostly stationary ice adhering to the headwall of the glacier valley.

At the outer border of the glacier, the ice melts away, and the rock material which was transported there, frozen into the ice, builds up a terminal moraine. Here develop conical hills called kames, deposited by the debouching meltwater streams, which also build alluvial fans of finer sediment in the surrounding area. In the distance (right), frost action produces a system of ice-wedges forming polygonal figures.

Plate 8

The same landscape as in Plate 7 after the melting of the glaciers. The valleys have the characteristic U-shape resulting from glacial action. The cirques now contain small lakes and the small cirque to the extreme left forms a so-called hanging valley where its stream of ice joined the larger glacier coming from the central cirque; small waterfalls descend from such hanging valleys. The lateral and medial moraines can still be seen, and the terminal moraine forms a distinct wall. The kames, marking the points where the meltwater streams debouched, rise above the terminal moraine, and the winding eskers going upstream from the kames show the course of the channels in the decaying ice sheet where the streams left much of their coarser gravel in their passage between the ice walls. Lakes mark the places where dead ice was left in the retreat stage, forming depressions in the ground. The small spoonshaped hills are drumlins, formed during temporary advances of the ice, their long axis being parallell to the direction of the ice-flow.

In the right distance, former frost polygons are still visible by their more luxuriant vegetation: the frost-cracks are now filled with sediment which is more fertile than the surrounding area.

Plate 9

During Rancholabrean glaciations, the withdrawal of the sea led to the emergence of a great land bridge between Siberia and Alaska, 600 miles wide. This was Beringia, a subcontinent in itself. Much of it was free of inland ice, and became populated by great herds of arctic mammals. Among those originating in the Old World, here are portrayed the steppe bison *(Bison priscus)* and the woolly mammoth *(Mammuthus primigenius)*. Both were extant on the Alaskan tundra as late as 12,000 years ago, and the reconstructions are based on frozen carcasses as well as European cave paintings depicting the same two species. Although this species of bison is distinct from the living American buffalo as well as from the living European wisent, both may be ultimately descended from it. The raised tail of the mammoth is shown in many cave paintings.

At peak glaciation, a barrier of ice made it impossible for animals to migrate southward from Alaska. When the climate ameliorated, during interglacial or interstadial spells, a migration route was opened. The first appearance of bison in America south of the ice heralds the beginning of the Rancholabrean age.

Plate 10

Only remnants of the great land fauna of ice-age Beringia are left, but the marine fauna survives nearly unchanged. Life in the sea along the Alaskan coast persists as it was off Beringia, and the two animals shown here—the harbor seal *(Phoca vitulina)* and the sea otter *(Enhydra lutris)*—were present during the last glaciation.

Plate 11

While the ice-sheets covered great tracts of northern North America, the extreme south was little affected by changes in climate. The main change resulted from the regression of the sea, which greatly increased the area of Florida. Then as now, the rivers were inhabited by the Florida manatee *(Trichechus manatus)*, seen here among mangrove roots. The fish are mangrove snappers.

Plate 12

The landscape and vegetation of a Florida river scene in the Pleistocene would look much as it does today, but the wildlife would be a strange mixture of the familiar and the exotic. The white ibis flying overhead *(Eudocimus albus)* still belong to the Florida scene, as do the palmetto and pond apple branches in the foreground, but the capybara has vanished. Of the two species of capybara present in Florida during the Wisconsinan glaciation, the painting shows the bigger, and rarer, form *Neochoerus pickneyi.* Its remains have been found at a few Floridian sites, and also in South Carolina, near Charleston.

Capybaras of the genus *Hydrochoerus,* also in the Florida Pleistocene, now live in tropical South and Central America. Their presence in Florida suggests that, during the Ice Age, winters in Florida were in fact warmer than they are now. This is one of many indications that the climate in southern North America was more equable in the Pleistocene than at the present day.

Plate 13

Mangrove, tapirs, and an anhinga bird drying its wings after a dive—this must have been a common scene on the coast of Florida during the Pleistocene. The tapir is *Tapirus veroensis*, a little larger than its modern South American congeners; the striped pattern on the young is also found in living species. Most finds of Vero tapir are from Florida, but it has also been recorded from Georgia, Kansas, Missouri, Tennessee, and Texas, and there are probable records from Kentucky, Pennsylvania, and Virginia. Such a northern distribution also suggests comparatively mild winters during the Pleistocene.

Plate 14 (*see next page*)

A fissure filling in limestone just south of Reddick, Marion County, has produced one of the greatest faunas of the Rancholabrean age in Florida —52 species of mammals, 64 species of birds, 32 species of reptitles, and 9 species of ampihibians, but no fish or other true aquatic forms. The fauna dates back to the Sangamonian interglacial (about 120,000 years), for the box turtle is a large lowland form suggesting high sea levels, and there are several typically interglacial transients from the south, including the ocelots *(Felis pardalis)* shown in the left foreground. Two species of bears occur at Reddick. The one to the right is the Florida cave bear *(Tremarctos floridanus),* a large, extinct relative of the living Andean bear of South America; it may well have had white face markings like the living form. The other is the black bear *(Ursus americanus),* a surviving species. Mastodon feces show that these big proboscideans, often found in Florida, have passed through the glade recently. In the background are two giant three-banded armadillos *(Holmesina septentrionalis).*

The fossiliferous sediment formed inside a cave where large amounts of rodent and snake bones were stockpiled by small predators, especially by owls. Bat, swallow, and vulture bones are also plentiful.

Plate 15

A group of stout-legged llamas *(Palaeolama mirifica)* is settling for the night in open ground not far from Reddick, Florida. The sun has set and the brief moment of twilight illuminates two of the llama's enemies: the great Pleistocene jaguar *(Panthera onca augusta)* and the vampire bat *(Desmodus stocki).*

Stout-legged llamas, originally of North American stock, evolved in the mountains of South and Central America, then re-immigrated to populate southern North America in a belt from California to Florida. Its main enemy may well have been the jaguar, Florida's most common predator in the Ice Age. Less deadly than the jaguar, the vampire is satisfied with a meal of blood, but may infect its victim with one of a number of dangerous diseases, including rabies.

The scene is set in the Sangamonian interglacial.

M. LAMBERT NEWMAN

Plate 16

From the "green winter" of Florida we return to the north, this time to the Great Plains, where this scene was enacted more than 34,000 years ago, near Hickman in western Kentucky. Returning from the Missouri River, a few miles away, a group of five flat-headed peccaries *(Platygonus compressus)*, heading east, is overtaken by a violent dust storm from the northwest. Unable to protect themselves from the dust, the animals will shortly perish and be covered by the accumulating wind-blown sediment.

Flat-headed peccaries were thought to be extinct, but a living member of the group, *Catagonus wagneri*, was recently discovered in South America.

Plate 17

Two short-faced bears, *Arctodus simus*, confront each other over the carcass of a horse *(Equus conversidens)*. The scene represents the southern Great Plains in Rancholabrean time. Greatest of all American predators in the Ice Age, this giant bear would have dwarfed even the redoubtable Kodiak bears of the present time. Related to the Andean bear rather than to grizzlies and black bears, it was re-markably long-limbed and slender of build for a bear, and probably capable of a burst of speed for a short distance somewhat in the manner of a lion.

Of the many horse species present in America in the Ice Age, the stout-legged *Equus conversidens* was one of the smallest, and evidently very common. Its size would be much like that of a present-day mule.

Plate 18

Flooded by fatigue after its violent attack, a scimitar-toothed cat *(Homotherium serum)*, one of a pair, stands over its prey, a baby woolly mammoth. Its mate, meanwhile, is drawing the enraged parent. Scenes like this probably occurred often in the Great Plains. In the north, as in this painting, the victim would be woolly mammoth *(Mammuthus primigenius)*; in the south, Jefferson mammoth *(M. jeffersonii)*, or perhaps the American mastodon *(Mammut americanum)*. A *Homotherium* den in the Friesenhahn Cave (Bexar County, Texas, near Bulverde) contained not only numerous skeletons of the cat and its cubs, but also milk teeth of hundreds of baby mammoths and mastodons—presumably the remains of prey which the cats had dragged to their lair.

The lion-sized scimitar cat has a long history in North America, going back to the Blancan; a related species lived in Europe in the late Pleistocene. Apart from Friesenhahn Cave, where the remains of perhaps 30 or more cats were found, fossils are rare, but the geographic range is wide, including Alaska, Florida, the Yukon, Oregon, Kansas, Tennessee, Oklahoma, and Texas.

The upper canine teeth of the scimitar cat were enlarged into long, recurved, stabbing structures. They were still larger in a second sabertoothed form, *Similodon fatalis* (See Plate 20).

Plate 19 (*see next page*)

One of the most remarkable Pleistocene sites in America is Rampart Cave, situated in northwestern Arizona in a cliff wall some 500 feet above the Colorado river. It takes us into the world of the shasta sloth, *Nothrotheriops shastensis*, for the floor of its 200-square feet chamber is covered by a thick deposit of sloth dung, preserved to our time because of the dryness of the climate. The deposit represents a time interval between 11,000 and more than 38,000 years B.P. and its bones and plant remains give a detailed history of the cave and its surroundings. The painting shows a shasta sloth feeding on prickly-pear cacti *(Opuntia)*; note the green tinge of the animal's fur, due to the presence of algae (remains of which were found in the fur of a mummified specimen from another site). Also shown are desert tortoise and Harrison's mountain goat *(Oreamnos harrisoni)*, an extinct species, smaller and shorter-legged than the living *Oreamnos americanus* of the Northwest. Plants include yucca, agave, and (in foreground) mormon tea *(Ephedra)*. Remains of these and several other animals and plants have been found in the cave.

Plate 20 (*see previous page*)

Aftermath of a sabertooth kill at Rancho La Brea, most famous of all Pleistocene sites in America. Situated in downtown Los Angeles, its tar-impregnated stream deposits have yielded the remains of 42 mammalian and 133 avian species as well as many other vertebrates, insects, plants, and so on. More than a thousand individuals of the great sabertooth *Smilodon fatalis californicus*, California's state fossil, are represented. The painting shows a resting female, whose huge canine tusks protrude well beyond the chin, with two cubs which still have small milk fangs. Several different kinds of vultures are now feeding on the carcass—they belong to the genera *Gymnogyps, Neogyps, Coragyps,* and *Cathartes;* the largest, on top, are *Teratornis,* which were among the greatest of all flying birds, but are now thought to have been of more predaceous bent than vultures in general. Patient storks *(Ciconia)* await their turn. Plants include cypress and (close-up) juniper. This amazingly prolific site is the type locality of the Rancholabrean age in North American Pleistocene history.

Plate 21

Tar-pool at Rancho la Brea, covered by a sheet of water, shimmers innocently while the morning mist is lifting. Stark remainders of the deadly trap underneath, bones of mammoth (left) and bison protrude out of the water. The mammoth is *Mammuthus jeffersonii;* the bison is *Bison antiquus,* thought to be ancestral to the living plains bison. A night heron *(Nycticorax nycticorax)* perches on the long dorsal spines of the bison's backbone, while a coyote *(Canis latrans)* forages on the bank and avocets *(Recurvirostra americana)* fly by. The birds and the coyote are extant today.

Plate 22

Ghostly in the mist, a pack of dire wolves travel past without noticing the brea pronghorn calf playing 'possum in the grasses; its watchful mother stands to the left. The dire wolf *(Canis dirus)* is more common in the brea pits than any other mammal species—remains of more than 1,500 individuals have been found. Its extremely powerful teeth and jaws suggest that it used to chew bones like a hyena. The brea pronghorn *(Capromeryx minor)*, whose forked horns gave it a four-horned appearance, was smaller than the living pronghorn antelope, and must have been one of the fastest-running mammals of all times. A small calf, however, would have been easy prey to the dire wolf. Foreground plants are blue elderberry; bird (at left) is cedar waxwing.

Plate 23

The San Josecito Cave in Nuevo León, Mexico, is notable for the enormous number of bones of the mountain deer *Navahoceros fricki* found in its deposits. It is situated more than 5,000 feet above sea level, near Aramberri in the southern part of the state. A total of 45 mammalian and 43 avian species have so far been identified in this great ossuary. The mountain deer, now extinct, is characterized by its very sturdy limbs and small, three-pronged antlers; it ranged along the Rocky Mountains into Wyoming. Its size was close to that of the living mule deer *(Odocoileus hemionus)*. A jaw and a limb bone of mountain deer are in foreground, also a badger *(Taxidea taxus)*.

Plate 24

Ground sloth *Glossotherium harlani* attempting to protect its young from dire wolf. These sloths could rely for defense not only on their powerfully clawed feet but also on a remarkably armored skin, reinforced by thousands of bone nodules forming a flexible mailcoat—this was revealed by the discovery of a dried skin of a closely related species in a South American cave. Harlan's ground sloth was a grasslands species, ranging widely in North America in the Rancholabrean, and the most common ground sloth in the tar pits of Rancho La Brea itself. Measuring more than four feet over the shoulders, and very massively built, it was second in size only to the gigantic *Eremotherium* among North American sloths. It had a keen sense of smell but the eyes were small and the tiny brain suggests a poorly developed intelligence.

Plage 25

The strange and the familiar blend in this Appalachian scene from the end of the last glaciation, showing mastodon *(Mammut americanum)* and white-tailed deer *(Odocoileus virginianus).* While mastodons were found in suitable environments over most of the continent, their real stronghold was eastern North America, where immense numbers of finds are known. Browsers, rather than grazers like the mammoths, they required wooded surroundings; they were equally at home in open spruce woodlands, spruce forests, hardwood groves, and swamps. Such a habitat would be shared with animals like the white-tailed deer, the wapiti, and the black bear.

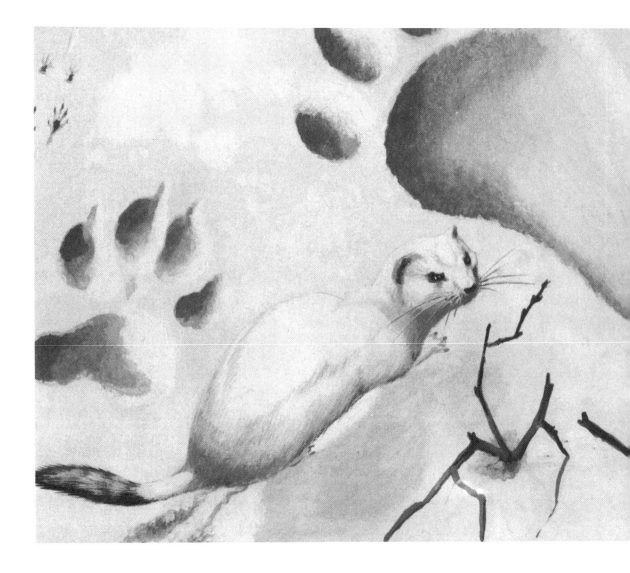

Plate 26

Many of the Appalachian caves contain a rich record of the life of the Ranchola-brean, and especially the small animals. In this scene, based on sites like the New Paris sinkhole (on the western flank of Chestnut Ridge in Bedford County, Pennsylvania), an ermine *(Mustela erminea)* pauses in the snow. Many animals have left their footprints: a black bear (center), a wolverine (left), a vole (far left, and another going into hole at right), a coyote (by the hole), and a wapiti (upper right). All of these species are still in existence.

M. LAMBERT NEWMAN

Plate 27

The time of the first appearance of man in America is still a controversial subject. There is evidence suggesting that it may have occurred as early as the Sangamonian (120,000 years ago) or even earlier, and at least one of the skulls found shows very primitive characters. However this may be—and future studies are eagerly awaited—it is evident that the entrance of the so-called Paleoindians, about 13,000 years ago, resulted in a great change in the American scene. The Paleoindians were anatomically modern men, ancestral to the Amerindians of the present day, and their hunting gear with exquisitely shaped, fluted projectile points indicates a specialization in big game. Early Paleoindians, the so-called Clovis people, hunted the mammoth and the mastodon, as shown by several finds of butchered specimens with scattered projectile points. Most finds are mammoth, but there are also sites where mastodons were killed; one of them is Kimmswick, south of St. Louis, Missouri, on which the scene in this painting is based. The mastodon was killed in a small pond, not far from the Mississippi River. Killing such a large animal was a big undertaking, calling for the combined efforts of several hunters. In this scene, a young man is calling for reinforcements, while a fellow hunter advances toward the wounded mastodon. The two young men to the left, who have thrown their spears, are now debating what to do next—such scenes of mingled confusion and determination may have been common.

Plate 28

After the demise of the great proboscideans (most mammoths and mastodons, perhaps all, were gone about 11,000 years ago) the Paleoindians concentrated their interest on the bison. These late Paleoindians, the Folsom people, developed a somewhat lighter projectile point. They would drive a bison herd, as in this painting into a suitable trap—such as the head of an arroyo, or the concave flank of a sand dune—where the animals could be killed off. Some of their bags were enormous, numbering a hundred heads or more, with an occasional camel *(Camelops hesternus)* among the bison. The bison herds consisted mainly of females and their young, with a few large males. Of 59 skulls from the Casper kill site in Wyoming, dated at 10,000 years ago, only three belonged to adult males, the remainder being females and young; there are even some fetuses. Bison hunting went on long after the Folsom cultural tradition had been abandoned—even the bison hunters at Casper belonged to a younger culture, named Hell Gap after another Wyoming site. At Hawken, also in Wyoming, a bison kill is dated at 6,500 years B.P. Yet bison remained plentiful in the Great Plains up to the time of indiscriminate slaughter in the nineteenth century.

V

The Green Winter

THE PLEISTOCENE IN FLORIDA

EUROPEAN SETTLERS coming to North America must have been struck by the unaccustomed richness and variety of the fauna and flora. For *one* species found in Europe, be it plant or animal, there are usually *many* equivalents in North America. And we can guess why. Just take a look at a map.

Before the Ice Age, the European fauna and flora were much richer in species than today. The Pleistocene brought about a weeding-out of species because almost all of the natural obstacles to migration run in a west-east direction: the Pyrenees, the Alps, the Mediterranean sea. As the ice sheets developed and the cold crept in from the north, animals and plants were pushed southward. Their final refuges were the Mediterranean peninsulas, Iberia, Italy, and Balkan. From these cul-de-sacs there was nowhere to go, and so many species became extinct. Indeed, it has been shown that each interglacial had a poorer flora than the preceding one.

In America the mountains run in a north-south direction and there is no real barrier until you come to the Gulf Coast. Even then, refuges were available in Mexico and Florida. So there was no weeding-out of species, and the fauna and flora persisted in all their richness and variety up to the end of the Ice Age. Since then, the fauna has been impoverished, but apparently from a very different cause.

One of the important refuge areas for warmth-loving species was Florida; and fortunately it so happens that the state has one of the richest Pleistocene records in the world.

Dr. S. David Webb, arriving in Florida from the arid West where fossiliferous strata erode before your eyes in the badlands, describes his sinking feeling as he encountered "soggy marshes, dense, deciduous forests, cypress swamps, and monotonous pine flatwoods" which he was supposed to probe for fossils. True, it is hard to see a fossil if the deposit is covered by vegetation. But he soon recognized that the swamps and marshes were ideal for the preservation of fossil remains, and that scuba diving in Florida's many waterways gives access to rich underwater sites. The drainage pattern of Florida seems to have changed little since the late Blancan, and so many fossiliferous stream deposits are located within or in the vicinity of present-day river beds.

Even more important for a study of Pleistocene life is the fact that the main part of the peninsula is underlain by a soft, easily soluble limestone. Sinkholes, caverns, and subterranean galleries riddle the bedrock like Swiss cheese. An entire lake may drain away, as if the plug had been ripped out of a bathtub. A river may vanish into a sinkhole, as happens with the Santa Fe River near the ghost town of O'Leno, only to emerge as a spring, perhaps miles away. Many of Florida's famous springs are well-known tourist attractions. Some are still untouched by man, and it is hoped will remain so, like the head-spring of Ichetucknee River, deep in the forest, with its dancing, lustrous, upwelling waters.

Pools and sinkholes form traps for animals whose remains become entombed in the bottom sediment. Mineralization may be remarkably rapid. Wilfred T. Neill tells of amusing examples, such as the remains of T-bone steaks being fully permineralized, and steel-sawed tree stumps hard as metal. The process is capricious: a lion jaw from Ichetucknee has one half dark-stained while the other half looks almost recent.

When the level of the groundwater sinks, caves and fissures become dry and may be inhabited by birds and beasts of prey which stockpile the remains of their catch. All this has made Florida a glorious hunting-ground for the student of Pleistocene life.

This has been long recognized. Joseph Leidy, a founding father of North American vertebrate paleontology, published on fossil mammals from Florida. Early in the twentieth century, the renowned student of early man in America, E. H. Sellards, worked in Florida; and for a few years around 1930, George Gaylord Simpson, the late dean of vertebrate paleontologists, turned his interest to the Floridian Pleistocene. The legacy of these great students has been well managed by a host of later workers. In 1974, Dave Webb listed more than thirty principal Pleistocene localities, the majority Rancholabrean in age.

A more precise dating of the sites is often difficult. Most of them—sinkholes, patchy stream and pool deposits, marshes and fissures—are isolated occurrences with no immediate clues as to age. Of two adjacent sinks, one may date from the Tertiary and the other from the Pleistocene. The fauna, of course, give a clue to the age, but only in a broad manner. How, for instance, to tell glacial and interglacial faunas apart, when the climate really did not change much? There are no arctic forms here: no musk-oxen, no caribou, no woolly mammoths.

Box turtle

The changes in water level may give an opening. Florida is a low-lying country with shoaling coasts, and so variations in sea level have a tremendous effect on the land area. The Wisconsinan fall of the sea more than doubled the size of the peninsula, the major increment falling on the Gulf side. It also led to a marked drop in the level of the ground water, which drains off through the fissure-riddled bedrock. As a result, there will often be a break in the sedimentation, so that the record of the low-water (or cold) intervals is largely missing. A warming-up with rising sea level will then produce back-filling and a particularly rich fossil record. Thus, we may expect a good coverage of the late Illinoian-early Sangamonian and, again, the late Wisconsinan-early Recent—the times when the northern ice-fields were on the wane.

Pursuing this line of reasoning, Walter Auffenberg, specialist on fossil reptiles, has worked out a most ingenious method. He uses fossil turtles.

The animal in question is the box turtle, *Terrapene carolina.* It belongs to the swamp turtles or emydids, but is exceptional in being fully terrestrial. It is a small turtle with a shell some five or six inches long, and ranges through the eastern United States from Florida up to the Canadian border. In the north, it escapes winter cold by digging itself down into the earth. It has a brown shell with beautiful yellow or orange markings. The face seems to bear an ingratiating smirk. Box turtles live on a variety of foods; the young ones are more partial to animal proteins, and the older are of a more vegetarian bent. You tell the sex by looking the animal in the eyes: the female iris is brown, but the males are fiercely red-eyed.

Auffenberg noted that there are two distinct subspecies of box turtle in the Florida Pleistocene (as, indeed, there are today). One, *T. carolina putnami,* is extinct; it is very large and seems to have inhabited coastal marsh and lowland savanna environments, as does its present-day successor, *T. carolina major.* The other subspecies, the living *T. carolina carolina,* occurs in upland forest. Hence, the presence of *putnami* suggests high waters and interglacial conditions; that of *carolina,* low waters and colder conditions. At one sinkhole by Haile in Alachua County, *carolina* is present in the lower deposits, *putnami* in the upper: we deduce a rising sea level. At Arredondo, also in Alachua County, the sequence is the reverse. Thus, the Haile site probably dates from the beginning of the Sangamonian interglacial, and the Arredondo one from its end.

The sites are mostly concentrated in the northern half of the peninsula. An especially rich fauna is known from a fissure filling just south of Reddick, Marion County. This truly prolific site has so far yielded 52 species of mammals, 64 of birds, 32 of reptiles, and 9 of amphibians. The box turtle here is the lowland *putnami*, in spite of the fact that the site is now 80 feet above sea level, so this was a time of interglacial high water—presumably Sangamonian. The impression is strengthened by the presence of tropical forms like a vampire bat, *Desmodus*, and the ocelot, *Felis pardalis*. Other probably Sangamonian sites are the Haile and Arredondo localities already mentioned and, for instance, Sabertooth Cave in Citrus County. The 27 mammalian species from here include not only the saber-toothed cat *(Smilodon)* but also a South American intruder, the little deer *Blastocerus*.

Many spring-run and river sites—Aucilla River in Jefferson County, Hornsby Springs in Alachua County, and Ichetucknee River in Columbia County—date from the Wisconsinan. Late Wisconsinan coastal marsh deposits are known in St. Petersburg, on the west coast, and Vero and Melbourne on the east coast. The Vero and Melbourne bone beds have long been famous for finds of human remains in the same deposit with Pleistocene fauna.

Another interesting site is a sinkhole trap in Levy County, called Devil's Den. The long list of species identified here includes several now-extinct Pleistocene forms together with the remains of man and domestic dogs. It is thought to date from well after the end of the Wisconsinan, and may record the survival of some of the last ice-age animals to exist in America.

In interglacial times, the environment in Florida probably was much like that of the present day, except that the higher water-level made the sea encroach upon the coastal plains. In glacial times, Florida retained its essentially subtropical climate, but the great increase of the land area probably brought about rather more continental conditions. A broad coastal plain was laid bare along the Gulf Coast. This may have formed a grassland or savanna "corridor" extending from southern Florida in a long sweep to the Caribbean coasts of Yucatán and Mexico. Such a great area, with a very uniform ecology, made it easy for southern forms to advance from Mexico to Florida. It may also have been a suitable environment to some of the Western grassland and desert species, and in fact a few typically western forms did spread into Florida. For instance, in Sabertooth Cave, the pocket gopher found is not *Geomys*, the eastern genus, but *Thomomys*, now an exclusively Western form. Again, at Haile, there is a ground squirrel of the Western genus *Spermophilus*.

The extraordinary variety of Pleistocene life in Florida reflects its crossroads position. Here, migrants from the tropical South and the arid West mingled with the fauna of the temperate regions to the north, retreating as the ice field extended into their habitats.

In Florida, more than anywhere else in America, land and water intermingle. From the lake-dotted interior to the mangrove archipelago of the coast, where the water takes over for good, the presence of waterways is always felt. Sometimes they are subterranean; sometimes they take the shape of a "river of grass" as in the marshes of the Everglades. So intimate is this relationship that it is only fitting to start a presentation of Pleistocene life in Florida with the animals that made up the "river community."

THE RIVER COMMUNITY

If you imagine a brown guinea pig measuring four feet from nose to tailless rump, and weighing about a hundred pounds, you have a pretty good picture of the animal called capybara. It is also called a water hog, and its Latin name *(Hydrochoerus hydrochoeris)* rather insists on it; but it is not a real hog. It is a rodent like the guinea pig, to which it is in fact related. It lives in and along the great river systems of the South American rain forests (a related but smaller species *Hydrochoerus isthmius,* the "isthmian water hog," is found in eastern Panama), and is the largest of all rodents living today.

Together with porcupines, guinea pigs, chinchillas, coypus, agoutis, and many others, the capybara belongs to the caviomorph rodents whose center of evolution was in the isolated continent of South America in Tertiary times. How their ancestors came in to begin with is still undecided. Most students believe they came from the north, by involuntary rafting; others think boldly in terms of a transatlantic crossing from Africa. At any rate the caviomorphs found excellent living conditions in their new homeland, and radiated into a great many kinds, large and small, just like the indigenous South Americans. Finally, when the land bridge was established, some of them found their way back to North America. Among these were the capybaras.

Although many caviomorphs are small, large forms are more common in this group than among other rodents: in it are found not only the largest living rodents but also the biggest rodents of all times. Some of them attained almost the size of a hippopotamus and probably lived in much the same way. A swimming capybara, too, is reminiscent of the hippo: it lies deep in the water, only the nose or top of the head are visible, and the nostrils are closed when diving. As in many other aquatic animals, the capybara has a layer of fat beneath its skin, and the fur, though long, is rather sparse.

Emerging out of the water, the capybara is revealed as a stocky, short-legged giant rodent with a large head. The head is very deep and has a conspicuously truncated nose, seen in profile; the eyes are situated high up so that the animal can look round when submerged while showing as little as possible of itself. Up on the river bank, the capybara may sit down just like a dog.

When in danger the capybara takes to the water and tries to hide beneath hanging tree branches or water-lily leaves. Its life is indeed far from charmed. Jaguars and caimans find it tasty, and so does man. Yet the animal seems to be quite successful.

The Pleistocene capybara in Florida was closely related to the living form, but it was slightly larger and sufficiently distinct to be regarded as a species of its own, *Hydrochoerus holmesi.* (It is named for Walter W. Holmes of the American Museum of Natural History, who was in charge of the excavations in Sabertooth Cave, Citrus County, where the first discovery was made in 1928.) Its record in Florida is long, for it is present in the early Irvingtonian of Inglis, and continued to flourish up to the end of the Pleistocene. So far, it has not been found outside the Peninsula, but it probably spread there along the Gulf Coast corridor in areas that are now beneath the sea. The fact that it existed in Florida during the northern glaciations suggests that the peninsular winters were actually milder than those of the present day.

Besides Holmes' capybara, a second and much larger species of the family was present in Florida. This is Pinckney's capybara, *Neochoerus pinckneyi.* As the distinct generic name implies, it was rather different from the living form. It was about six feet long and had a broad, flat-topped head. It ranged more widely than *Hydrochoerus,* to southern Texas along the Gulf Coast and to Charleston, South Carolina, along the Atlantic. All of the finds are Rancholabrean in age and the origin of *Neochoerus* is unknown.

Capybaras, though fond of water, are amphibious and move easily on the ground. The same is true for the tapir, another common member of the river community of Ice Age Florida. Like capybaras, tapirs are now typical of the Latin American fauna. But their origin is in the north and they are not rodents but ungulates, related to the horses.

At present, three species of tapirs live in Latin America, ranging north to southern Mexico; and a fourth lives in southeastern Asia. They are rather similar, including the Asian form whose black-and-white coloring contrasts with the mainly self-colored, dark brown to greyish black American species. Under the skin they are all much the same, and are all placed in the single genus *Tapirus.* They are remnants of a once-continuous tapir population of North America and Eurasia in the Tertiary, and the South American forms are late invaders from the north.

Tapirs are evolutionary conservatives, and the stock can be traced back some fifty million years. The only notable specialization of these archaic mammals is the presence of a small proboscis. Dangling, it gives the animal that inimitable tapir look; when stretched forward in sniffing, the tapir seems suddenly transformed into a long-snouted pig.

Most tapirs are lowland forms, but one living species, the rare mountain tapir, lives in the Andes of northern South America.

Amphibious mammals tend often to be large and heavy, and this is true for the tapir too; they may weigh up to 500 or 600 pounds. Their fur is short and dense, and the skin very thick. This is very useful, for a frightened tapir will seek the densest scrub at hand and burst through it unscathed, so escaping more sensitive-skinned pursuers, like the jaguar. In the open, on the other hand, it is easily hunted down.

But mostly the tapir keeps close to the water and, like capybaras, tries to escape by swimming.

The Pleistocene tapirs of North America are only found in the unglaciated areas, and only in such regions where the annual rainfall now exceeds 500 millimeters: along the Pacific coast and east of the 100-degree meridian. The western forms are little known, but in the east the fossil material is abundant and we know that Florida was inhabited by two species. The commonest one, the Vero tapir *(Tapirus veroensis)* is but slightly larger than the living South American ones; it is only known from Rancholabrean times. Sparser finds record the presence of a much larger species, Cope's tapir *(Tapirus copei)* from the early Irvingtonian on.

Although perfectly able to move on land, tapirs and capybaras must certainly be reckoned members of the river community of Florida; and the same is, of course, true for such smaller creatures as beavers, water rats, muskrats, and so on (the muskrat is now gone in Florida). But there was, and still is, another member of this community, an inoffensive plant-feeder like the others, but unable to leave the water: the Florida manatee, *Trichechus manatus.*

The manatee, a ten-foot aquatic mammal, belongs to the order Sirenia, like the sea-cow described in the previous chapter. Manatees, however, are reckoned a separate family within the order. Externally, they may be identified from their rounded tail flukes; and a dissection will show that they, for reasons unknown, have only six vertebrae in their necks, instead of seven as in other mammals. They have very thick ribs, which fossilize easily. The bone structure is in fact very dense, like ivory, and reminiscent of a pathological condition termed *pachyostosis* in man.

There are only three living species of manatees in the world. *Trichechus manatus* is one. A second species lives in the Amazon and Orinoco river systems; the third in western Africa. Unlike the other species, which prefer freshwater, the Florida manatee lives in the sea as well as in rivers, and ranges north to the coast of North Carolina. The Pleistocene range was about the same: the northernmost occurrence, dating from Sangamonian times, is at Fort Fisher in New Hanover County, N.C. The species seems to be precariously close to its northern limit, even in Florida. It suffers badly from cold spells and several manatees can then be seen to huddle together. Or they may seek a spring with constant water temperature, in spite of the water being much too clear for their liking (they prefer it muddy). In the crystal spring water, the manatees, probably feeling uncomfortable, may be seen promenading around on the bottom, using their long flippers as legs. When the weather warms up, they return to their favorite haunts to browse on water hyacinths, water lilies, and other floating plants. They also nibble plants from the shore, and then have to stand in the water with their heads in the air.

Unlike most others of the great Pleistocene beasts, the manatee has survived to the present day. Hunting, in combination with severe winters like that of 1940, has been a serious threat to their existence. Conservation measures are necessary for the protection of this remarkable if not exactly beautiful creature. Some dog-lovers say about some dog-breeds, "It's so ugly it's sweet." Some of us may feel that the same applies to *Trichechus manatus.* After all, it has been around here since Blancan times, and we don't want to lose it.

THE REPTILIAN RULERS

The ruler of Florida's waterways, now as in the Ice Age, is *Alligator mississipiensis,* the American alligator. The name "alligator" is a corrupted version of *el lagarto,* Spanish for "the lizard," since elevated to the status of a scientific generic name. The trivial name conserves the original spelling of the Mississippi, just as *Homo neanderthalensis* conserves the pre-1905 spelling of Neandertal (biological terminology, too, has its fossils).

The cold never really touched Florida, and so the alligator is common in the Pleistocene deposits here. In historical times it ranged from the mouth of the Rio Grande in southern Texas to Dismal Swamp in Virginia and North Carolina. It is an oldtimer in Florida. In Miocene times, fifteen to twenty million years ago, there lived an ancestral species, *Alligator olseni,* which was smaller than its descendant. By Blancan times it had given rise to the modern form. But the story of *Alligator* goes still further back and experts now think the genus originated in North America well back in the Oligocene, maybe thirty million years ago, a time when swamplands were widespread over the continent, and the climate much warmer than today. One branch of the genus reached Asia and gave rise to the Chinese alligator, the other living species of the genus. The caimans of South America, although alligatorine crocodilians, have a separate history going back to early Tertiary times.

Like caimans, gavials, and crocodiles, alligators are members of the order Crocodilia. The order is very old, over 200 million years. Together with the crocodilians, and from the same ancestral stock, arose three other reptilian orders which are now extinct: two orders of dinosaurs, and one of flying reptiles. Today's crocodilians are the only living representatives of this great array of "ruling reptiles."

A big alligator is indeed an impressive sight, and a reminder of that Age of Reptiles long ago. No wonder that many myths are circulated about the alligator—not only in the backwoods, but also in some "scientific" writings. Wilfred T. Neill, Florida's outstanding student of the Crocodilia, makes short shrift of them. Here are some items:

Only male alligators roar. (Females do, just as lustily.)
Alligators may become several hundred years old. (They lose most of their teeth before fifty, and are shorter-lived than man.)

Some alligators are so old that moss grows on their backs. (The growth is algal and can appear quickly on a sick alligator.)

Only the tail is edible. (Alligator meat is edible throughout and the tail is not the best.)

Alligators use their tails when attacking and fighting. (They do not.)

Alligators bask in the sun with their jaws open. (They prefer the shade and keep their mouths closed, but open the jaws if you come close.)

Alligators lay hundreds of eggs in a tiered nest. (The total is 30–50 and they are laid in a cluster.)

In fact most of the early literature on alligators is legendary, and so much of the mythology has persisted in later writings that the life and habits of this remarkable animal are much less known than one might imagine.

Alligators are animals of the coastal lowlands where they may be found in situations ranging from tree-shaded rivers to open marshes studded with water holes. They favor shallow waters where they keep close to the surface. They also range into brackish coastal waters but not into the open sea, and the coast is the area where the alligator cedes his place to his distant cousin, the American crocodile.

The life history of an alligator begins in the spring with the courtship of the putative parents, which is a drawn-out affair ending with copulation in a side-by-side position. The eggs are laid in a nest which the female makes out of dead leaves and mud or sand well up from the water's edge. The nest is built in the shade (for instance of a large tree), for the main danger to the eggs is overheating; and the female will guard the nest and drive off intruders (raccoons, for instance, are keen on alligator eggs).

When the eggs are hatched, the female will help to dig out the hatchlings, and may carry them down to the water in her mouth. (This behavior is also seen in true crocodiles like the Nile crocodile of Africa, and so presumably evolved at an early stage in crocodilian evolution, for alligators and crocodiles have been separate since Cretaceous times.) The young alligators then scatter rapidly along the waterways. At that time in their life, they are heavily preyed upon, their enemies being the blue heron and the long-legged caracara hawk; even the larger frogs may catch small alligators. But for those few who survive all the dangers, the situation is reversed as the animal enters its mature estate.

The alligator mainly hunts vertebrates. Fish, aquatic turtles and snakes, water birds, rodents, marsh rabbits, and minks figure in its menu. Large alligators may take bigger prey such as dogs, calves, or even cows, and occasionally are known to attack man, although this is much less common than in the case of the true crocodiles. In these days the alligator is hunted so relentlessly that few get the chance to grow to really large size. In the Pleistocene the situation was different and the alligator had plenty of large mammals as potential prey—perhaps mainly its neighbors in the

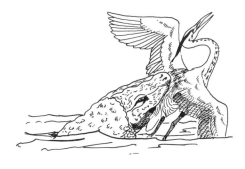

Heron eats gator—gator eats heron. A new version of the old puzzle, "Which came first, the chicken or the egg?"

river community, the capybara (which is the favorite prey of the South American caiman) and the tapir.

Contrary to legend, alligators are not cannibals. Juvenile alligators are mostly ignored by the adults, except if they cry out when seized by an enemy; that distress call will stimulate every adult alligator to rush to the place and attack the predator.

Few if any animals are able to prey on adult alligators. Although bears and pumas (and the great Pleistocene jaguar) could perhaps do so, this has not been confirmed. We are probably justified in seeing *Alligator mississipiensis* as the undisputed ruler of Florida's Pleistocene river community.

In a way, the rule was a benign one. Alligators are cold-blooded animals and do not need much food. A mammalian carnivore of the same size would eat at least ten times as much. Also, alligators are inactive during the winter. In fact, winter cold was a main factor limiting the range of the species—as in the case of the manatees.

The situation has now changed. Man is the alligator's principal enemy and the main limiting factor. This is due not only to hunting but also to destruction of the environment. This sometimes occurs in a subtle way but it nevertheless threatens the survival of the original inhabitants. If alligators are to survive, their ancient life zone must also be preserved. That kingly species, *Alligator mississipiensis*, needs all the friends it can get.

Where the fresh waters draining off Florida finally mingle with the salt waters of the sea, another and rather different species of crocodilian takes over. This is the American crocodile, *Crocodylus acutus.* It is easily distinguished from the blunt-snouted alligator by its long, narrow jaws and by its teeth, which are visible also when the mouth is closed. To a human observer, this makes it seem a much bloodthirstier beast than the alligator (and has given it such a reputation, too), although it is actually quite shy.

Crocodylus is one of the oldest reptilian genera in existence: it arose toward the end of the Age of Reptiles more than 65 million years ago, probably in Asia. The first New World *Crocodylus* are known from the Eocene, about 50 million years ago.

The range of the American crocodile reaches only the southern tip of Florida and encompasses Florida Bay and the coast up to Cape Sable, although in historical times it ranged to Palm Beach on the east coast and Charlotte Harbor on the west. Very likely it ranged to Florida in the Pleistocene too, at least in interglacial times; but there are no certain fossil occurrences. Deposits in which remains may exist are probably submerged at present. This species keeps to the sea and ascends only the largest rivers; and there is no Florida river big enough for it. And so, in Florida, the crocodile is not a member of the river community, although it may well have been so elsewhere.

THE ARMORED ONES

In no part of the United States were the Edentata, those invaders from the south, more varied and numerous than in the Southeast. Ground sloths, great tortoiselike glyptodonts, and armadillos—all were present in Florida, and all vanished at the end of the Pleistocene.

Today, there are wild armadillos in Florida once more. Present only in southernmost Texas around 1850, the nine-banded armadillo, *Dasypus novemcinctus,* has since spread far to the north and east—with some help from man. Once, it even reached southern Nebraska, but increasing winter cold drove it back and it is no longer found north of Kansas. But in Florida, where it was introduced by man, it is happily extending its territory. Once more, Florida is becoming edentate country.

This intriguing little animal is about the size of a cat, but decidedly not as soft to the touch. Its body is encased in a jointed carapace consisting of one shield in front and another in the back, with nine (or sometimes eight, or up to eleven) bands in between, movable against each other so that it can roll up into a little ball. The tail has a tubular sheath. The armor is made up of bony scutes covered by a horny shield; after death the horny substance disintegrates and the scutes may be scattered about and get entombed in the sediment. As one individual will produce hundreds of scutes, chances are obviously good for recovering a record of its presence.

The only adornment of this long-nosed little creature is a pair of incongruously-

Carrying its armor lightly: nine-banded armadillo *(Dasypus novemcinctus),* making Florida
edentate land once more.

long, inquisitive ears which seem somehow out of place in a digging animal. For the
armadillo is powerful digger indeed. It always keeps to soft ground, where it digs its
lair, lining it with leaves (which are renewed when the bedding gets moldy). Not satisfied with one lair, it usually digs several other tunnels and dens, for unknown reasons—perhaps just to get a change of scene.

The armadillo feeds on worms, grubs, and insects, which it digs out of the ground
after locating them with its very sensitive nose. Very occasionally it goes carnivorous
on a somewhat grander scale by killing newborn rabbits, for instance, though chew-
ing them up seems to be a tough job. An armadillo on the hunt is easily located because of its incessant loud snorting, which has been likened to the noise of some
wind-up toy. This seems reckless, as there may be predators about, but in fact the
intrepid little beast is well protected by its armor and is able to dig itself down into
the ground so fast that its enemy will usually be baffled. It can swim, too, in spite of
its heavy armor; it becomes buoyant by filling not only its lungs, but also its gut, with
air. So nature has endowed this little (and rather stupid) animal with an amazing
versatility.

There are no Pleistocene fossils of the nine-banded armadillo. The earliest remains, from Miller's Cave in Llano County, Texas, are only about 3,000 years old. The
Pleistocene armadillo was another species about twice the size of the living one: *Dasypus bellus,* the beautiful armadillo. It was an old-timer in Florida, where it appeared in Blancan times and continued to live throughout the Pleistocene. A remarkable find of a complete female skeleton with unborn embryos inside its carapace, from
a cave near Reddick, shows us an animal about four feet in length. The Blancan ancestor
was a bit smaller. In Irvingtonian and Rancholabrean times, the beautiful armadillo
spread into Texas, Missouri, and West Virginia, and at least in Texas (Miller's Cave
once more, but an older level) it survived well beyond the end of the Wisconsinan: the
date is 7,300 B.P. Apart from size, *Dasypus bellus* probably was closely similar to the
living species, and lived in the same manner.

A four-foot armadillo seems quite an impressive creature, though it is eclipsed by
the living giant armadillo of South America, with a total length of about seven feet.
But even that one is dwarfed by the real giants of the family, the Pleistocene pampatheres. These amazing beasts reached almost the size of a small rhinoceros. The North

American form, *Holmesina septentrionalis*, has been found in Florida, Texas, Oklahoma, and Kansas; most finds are isolated scutes and limb bones, but there is a good skeleton from Houston, Texas. It shows that there were only three movable bands between the anterior and posterior shields. The tail was protected by a rigid tube. Again, this is a Florida old-timer, descended from a smaller Blancan form.

We may speculate about the habits of pampatheres. They were probably too large to be such prolific diggers as the living armadillos, but otherwise, with their similar skeletons and long snouts, they resemble *Dasypus* so much that we may suspect they were insect-eaters too. Some armadillos do take plant food as well, and the large ones may have been (at least part-time) vegetarians.

This seems certainly true for the real giants among the armored edentates: the wholly extinct glyptodonts. Although related to armadillos, and ultimately derived from them, these amazing mammalian tortoises have a long separate history in South America. Like their relatives, they entered North America in Blancan times, and reached Texas and Arizona. They entered Florida in the Irvingtonian and remained a prominent element in its fauna to the end of the Pleistocene. They seem to have inhabited the coastal plains of Florida and Texas, and the Atlantic seaboard as far north as South Carolina. All of the North American forms belong to a single genus, *Glyptotherium*, and apparently represent a single lineage with three successive species: *texanum* in the Blancan; *arizonae* in the Irvingtonian; and *floridanum* in the Rancholabrean. Size culminated in the Irvingtonian species, but some *floridanum* males may have been as big as *arizonae*.

A glyptodont looked like nothing on earth today. The nearest approach to it might be a giant land tortoise, but the glyptodont had a longer tail and, of course, was a mammal, not a reptile. It was encased in a high-vaulted carapace covering the entire body and reaching a length of up to six feet. The animal as a whole, with the tail, was up to ten feet long, with a standing height of up to five feet. It weighed a ton or more. The carapace was formed by nearly two thousand body scutes welded together into a single, nearly inflexible shield. In the early history of the glyptodonts in South America, we can see this structure evolving out of an earlier model with movable bands, as in the armadillo. Apparently the protection afforded by the undivided shield was more important to the glyptodonts than the flexibility of the earlier type. The head was protected by a casque of the same construction, and the tail was surrounded by rings of armor forming a long tube.

The limbs were pillarlike, as in elephants, but very short, and the claws (which are prominent in other edentates) had been modified into hooflike nails. The animal must have moved very slowly and clumsily. To complete the weird picture, David Gillette finds that the glyptodonts probably had a small proboscis!

Can we visualize the mode of life of animals so utterly unlike anything living today? At least we can try. They lived, almost universally, in moist and warm coastal areas and along river banks. They were generally found together with capybara, and so probably spent much of their lives in or near water. With their ponderous bodies and awkward movements, they were surely exclusive plant-feeders like today's big land tortoises. And for protection against enemies, they relied mainly on their armor. But they probably used their tails too, club fashion. A South American glyptodont of another genus *(Daedicurus)* carried at the end of its tail a veritable mace, consisting of long bony spikes: this was probably a defensive weapon.

The glyptodonts may not always have succeeded in defending themselves. A skull of a Blancan glyptodont has two punctures which evidently were made by the canine teeth of a great cat. An almost allegorical scene rises to the mind. They had evolved for millions of years, half a world apart: the armored, peanut-brained superdreadnought and the supple, intelligent, toothy carnivore. Here at last they met on the Gulf Coast and took stock of each other. The lesson is that of the "arms race." Even the heaviest armor will eventually find the gun that packs a sufficient punch.

THE UPLAND COMMUNITY

The glyptodonts, in a way, might be reckoned members of the river community. The main habitats of the armadillos, on the other hand, were probably the upland savannas and forests of Florida. The upland community was even more varied than that of the streams. Sloths, mastodons, mammoths, horses, peccaries, deer, pronghorns, bison, wolves, bears, and great cats lived there, as well as a great host of smaller mammals, birds, lower vertebrates, and other organisms. Here, we can only take a look at a few of its representatives, and choose those that were especially characteristic or common in the area.

The edentates of North America, besides the armored ones, include the great sloths, also of South American origin. Three families of ground sloths lived in North America in the Pleistocene, and all were present in Florida. The Mylodontidae were represented by *Glossotherium*, and the Megalonychidae by *Megalonyx*. They will be treated in later chapters, for they were widespread forms and the Florida occurrence is marginal. But the biggest of them all, *Eremotherium rusconii* of the Megatheriidae, has only been found in the Southeast, and most of the sites are in Florida; in addition, there are a few occurrences in Georgia, South Carolina, and Texas. A single site (Inglis) dates from the Irvingtonian, but all the others are Rancholabrean in age. The species may have spread twice into the Southeast along the Gulf Coast corridor from Central America, where there probably was a standing population. Other species of the genus *Eremotherium* lived in South America.

The name comes from the Greek *eremos*, or hermit, and the vernacular "hermit sloth" may be suggested. In the past, the genus was confused with the related South American *Megatherium*, which was equally gigantic. The two actually differ in many anatomical details; for instance, the forefoot of the hermit sloth has only three digits, that of the megathere four.

Finds of hermit sloth in the United States are fragmentary: most are isolated teeth or limb bones. Fortunately, well preserved material has been unearthed in Panama, so that we can form an image of this extraordinary beast.

Eremotherium rusconii reached a total length of up to twenty feet from the nose to the tip of the massive tail, and was the only Rancholabrean land mammal to compete with mammoths and mastodons in weight. The hind limbs were immensely stout and the feet very big and flat. While *Megatherium* walked on the outer edges of the feet, the foot of *Eremotherium* rested upon its heel and sole. The hind feet and the tail formed a firm tripod which enabled the hermit sloth to rear up and browse from the trees, perhaps using its clawed forefeet to pull down the branches. It was thus a treetop browser like the giraffe, but attained this specialization in a completely different way. We have no direct evidence of the food of the Florida hermit sloth, but there are remains of twigs associated with eremothere remains in Peru and Ecuador which are chopped up in lengths corresponding to the interspaces between the crests of its cheek teeth. The plant remains suggest that the animals inhabited savannas with thorny bushes.

When moving about, the hermit sloth would amble along on all fours, walking on the knuckles of the front feet with the two claws turned backward (one of the fingers was clawless). It could only have moved slowly and awkwardly. Its main protection was its immense size and strength, but the claws of the front feet may have been used as weapons of defence. Like other edentates, it had a very small brain and presumably was an unintelligent creature.

Another typically Floridian herbivore, probably with about the same geographic distribution and environmental preferences as the hermit sloth, was the stout-legged llama, *Palaeolama mirifica*. Finds of this animal, too, are concentrated in Florida, from the Irvingtonian to the end of the Rancholabrean. But in this case, additional specimens come not only from Texas but also from California. The genus probably originated in the Andean region of South America, from which it spread to Peru, Ecuador, Central America, and then northward to California and eastward along the Gulf Coast. Its very stocky limb bones suggest that it arose originally as a mountain form. The nature of Florida, although wonderful, does not include any alpine habitats, so we must assume a secondary adaptation, perhaps to a savanna environment where *Palaeolama* fed on grass and on the shoots and leaves of trees and bushes.

Palaeolama seems to be a specialized descendant of the widespread genus *Hemiauchenia*, which, as we have seen, was common in the southern part of North America since Blancan times; *Hemiauchenia*, too, was present in Florida. The two were quite dissimilar, for *Hemiauchenia* had long, slender limbs and must have been very fleet of foot. Their modes of life must have differed greatly and this made it possible for them to exist in the same general area without being thrown into competition. Probably they had slightly different environmental preferences, for among the 27 sites in Florida where one or the other have been found, there are only three in which they occurred together.

We may note that even such a small and seemingly homogeneous group of animals as the llamas has had a very complex history. At present, four South American species survive, two in the wild (the guanaco and the vicuña), and two as domesticated animals (the llama and the alpaca).

Leaving the herbivores aside for the moment, we will turn to the flesh-eaters that preyed on them. Among them were two for which Florida was a particular stronghold in the Pleistocene: the Florida cave bear and the jaguar.

Of the seven living species of bears, the least-known, as well as the smallest, is the South American "spectacled" bear, *Tremarctos ornatus*, hiding in his wooded haunts on the slopes of the Andes. The fur is soot-black, but there is some white and brown on the nose and throat, and the yellowish stripes above the eyes are somewhat suggestive of spectacle bows. There is something lissome and almost doglike about the build of this bear which sets it apart from all the other living species, and in fact it is only distantly related to them.

The Andean bear is the last scion of a once-mighty stock of all-American bears. The story of the tremarctines may go back some ten million years, and in that time they produced many wonderful forms, some of which were probably the top predators of their world.

The Florida cave bear, *Tremarctos floridanus*, was closely related to the Andean bear, but much larger. I estimate the weight of a large male at about 600 pounds, or four times that of its living cousin. But apparently he was not particularly predaceous in habits. He is more likely to have had a vegetarian bent, like the European cave bear whom he seems to mimic. We will digress with a few words on that species.

During the Ice Age, Europe was inhabited by a very large species of bear, *Ursus spelaeus* or the cave bear, closely related to the brown and grizzly bears (species *Ursus arctos*). We have no direct evidence on its habits, but compared with the grizzly it had much larger and blunter-cusped molars—apparently an adaptation to vegetarian food. Albert Gaudry, with the Frenchman's gift for the expressive phrase, called it "the least carnivorous of carnivores, and the most bearish of the bears."

It was my privilege to study the remains of *Tremarctos floridanus* and it was amazing to see them depart from *Tremarctos ornatus* in just the same way as *Ursus spelaeus* departed from *Ursus arctos*. The Florida cave bear, too, was much larger than its congener. In both forms, males are much bigger than females. In both, the back teeth are expanded and blunt-cusped. And in both, the skull was domed to give good leverage for the muscles acting on the molars. The shortened feet and extremely heavy limbs told the same story. The two cave bear species, the American and the European, are a good example of parallel evolution, resulting from adaptation to a similar mode of life.

As we shall see in the next chapter, the tremarctine stock also gave rise to a great predator like the grizzly; but that one is rare or absent in Florida.

Most of the remains of *Tremarctos floridanus* come from sinkhole and river deposits in Florida, but the species has been found elsewhere too. In Blancan times it ranged north to Idaho, but in the Rancholabrean it was restricted to the south, ranging from Mexico and New Mexico along the Gulf Coast, and into Tennessee and Georgia. Numerous skeletons from the Devil's Den sinkhole in Levy County show that this bear survived into the early Recent, when most of the typical Pleistocene mammals were extinct.

Remains of the cave bear are often found in association with those of the extant black bear, *Ursus americanus,* which still lives in Florida. Indeed, if there are many fossils of *Tremarctos,* it is a safe bet that the black bear will be present too. The Pleistocene black bear was larger than the living one, about the size of a small grizzly, but smaller than the cave bear. (Oddly enough, the present-day black bear of Florida averages distinctly larger than elsewhere.) The prolonged coexistence of the two species of bears suggests that they were not in direct competition, and somehow had managed to parcel out the "bear niche" between them. Perhaps the black bear was the more carnivorous of the two.

The top carnivore in most of Florida, however, may have been the jaguar. Lion remains are rare and restricted to northern Florida. When the lion arrived here, in the late Rancholabrean, the jaguar was already ensconced and probably was better adapted to the forests, marshlands, and streams of Florida.

The Pleistocene jaguar was much larger than the one living now, and this is especially true for the huge Irvingtonian jaguar. In the course of the Pleistocene, jaguars tended to become somewhat smaller, and their limbs became shorter and stockier. The first jaguars in America were probably somewhat lionlike, but as they adapted to the environment in the south with dense forests, numerous small streams and, in many cases, rocky and broken terrain, they became more like the living form. Here is an excellent instance of gradual change within an evolving lineage, with small cumulative steps leading to an end form so unlike the first invader that the two can be regarded as distinct species.

Although not as huge as their Irvingtonian ancestors, the late Rancholabrean jaguars were still much larger than any living ones, so that the main size reduction must have occurred after the end of the Wisconsinan. A few jaguar bones from the early Recent Devil's Den sinkhole give a glimpse of this process—they are smaller than the Pleistocene ones, yet bigger than modern jaguar. There are many other instances of such reduction in size, and the phenomenon will be discussed in more detail later on.

The commonest great cat besides the jaguar is the sabertooth *Smilodon,* whereas the scimitar-toothed *Homotherium* is rare as always. Both will be treated in later chapters, as will the coyotes and wolves. The puma, often called "panther" in Florida, was present in the Pleistocene too. Among the smaller cats, the bobcat is frequently found, while exotic forms like jaguarundi and ocelot appear occasionally. Other small carnivores, not very common in the deposits, include foxes (mainly the grey fox), raccoons, weasels, and skunks.

The Pleistocene bird fauna of Florida, though only fractionally preserved, is overwhelmingly rich; we owe our knowledge of it to the lifework of Pierce Brodkorb of

Gainesville. The interglacial site of Reddick, for instance, has yielded over fifty species, of which about one-fifth are extinct. The grebes, ducks, raptors, quails, turkeys, rails, killdeer, snipe, pigeons, owls, woodpeckers, kingbirds, swallows, crows, wrens, grackles, and finches in this fauna were sampled from a variety of surrounding habitats. The same deposit contains the remains of twenty species of snakes and five lizards, all of them extant. The turtles are box turtles, gophers, soft-shelled turtles, and two extinct species of the land tortoise, genus *Geochelone*. Frogs, toads, and salamanders, a total of nine species, make up the rest of the vertebrates; only fishes are missing.

With most of the large mammals gone, the greater part of the Pleistocene wildlife in Florida is still in existence. Long may this continue.

VI
Windswept Plains

THE SETTING

THE GREAT PLAINS and Central Lowland form, roughly, an isosceles triangle, its base extending from the Edwards Plateau of southern Texas to Lake Ontario, and its apex far to the northwest in the area of Great Bear Lake in Canada's Northwest Territories. Rain-shadow from the Rockies robs much of its western part of moisture; farther east more humid conditions prevail. So there is a transition from desert to short-grass prairie to tall-grass prairie to forest.

In Blancan times, conditions were moister throughout and forests grew in many areas which are now treeless. Irvingtonian climates brought increasing dryness and glacial-interglacial cycles of cold and warm; and the cycles were intensified in the Rancholabrean. In the present interglacial, this vast area accommodates a spectrum of environments from arctic tundra to southern desert. At the time of the Wisconsinan glacial maximum, one-half or more of it was covered by ice, as proved by the distribution of glacial tills and moraines, and south of the ice-margin the climatic belts were squeezed and telescoped correspondingly. What has not been recognized until recently is that the climatic belts did not just shift southward and become compressed; they also changed their very nature, sometimes beyond recognition. We are barely beginning to understand the ice-age conditions in the unglaciated parts of the Great Plains.

The Wisconsinan land ice on the Atlantic seaboard extended just south of Long Island. The Island itself and the coast to the northeast were under the ice, which extended onto the continental shelf and presented a high ice-cliff toward the sea. Along the ice-front flowed the warm Florida Gulf Stream, and the conditions must have been singularly unpleasant, with storms, thick fogs, and floating icebergs calved off the cliff.

From New York the ice-margin extended westward, skirting most of Pennsylvania, but forming a great southward-extending lobe through Ohio and Indiana, following in part the course of the Ohio River. In Illinois it turned north, leaving the southwestern corner of Wisconsin ice-free (the so-called "driftless area"), but then turned south to form another, smaller lobe extending into Iowa. The margin can then be traced to the northwest through South and North Dakota, approximately along the Missouri, and from Montana on it undulated westward more or less along the 48th parallel to Puget Sound. There were prominent southern extensions and numerous separate icefields along the Rocky Mountains and the Cascade Range. Great ice sheets formed as far south as the Sierra Nevada and the Colorado Rockies.

In some regions of the Great Plains, the Wisconsinan ice did not extend as far as some of the earlier glaciations. In Illinoian times, for instance, the Indiana lobe extended far to the southwest, overrunning most of Illinois; and the Kansan ice sheet invaded the northeastern corner of Kansas.

South of the ice, the landscape probably was a treeless tundra. But this steppe-tundra of the middle latitudes must have been rather different from the arctic tundra of today with its dark winters and light summers, and some of its animals—the peccaries, for instance—seem curiously out of place. As we go farther south, even more paradoxical features become apparent. It almost seems as if the winters in the southern Great Plains were milder than at the present time, and the climate more equable. It is true that the "astronomical theory" of the Ice Ages predicts milder winters and cooler summers during glacial intervals.

As northern species were displaced and forced southward by the ice sheet, we might expect an "accordion effect," the resident species becoming displaced to the south as well. But such was not the case. For instance, the resident voles and shrews, from Texas to southern Kansas, Arkansas, and Missouri, remained in place together

Among exclusively American rodents, pocket gophers have a good fossil history, probably owing to their burrowing habits. More than 20 species are known from the Blancan and Pleistocene.

with such northern invaders as the boreal redback vole, the meadow vole, and the arctic shrew. The resulting animal communities thus were unlike those of the present day, even when we consider only those species that are still in existence. Several workers, for instance Russell W. Graham at Purdue and Ernest L. Lundelius at the University of Texas, have paid attention to this interesting phenomenon in recent years.

Lundelius classifies the Wisconsinan mammals of central Texas into three groups. The first consists of the great extinct forms: the mammoths and mastodons, the big sloths and glyptodonts, the great bison, the horses and peccaries, camels and llamas, and such large carnivores as the lion, the saber-toothed cats, the short-faced bears, and the dire wolves. A second group consists of living species which are no more found in this area, but exist farther to the north: small forms like the masked and shorttail shrews, the ermine, the prairie vole, the muskrat, and the bog lemming. In the third group, finally, are the species which still live in the area: for instance, the white-tailed deer, coyote, wolf, bobcat, and spotted and striped skunks.

The second group points to a more humid climate than that of today. It has long been known that the glacial episodes in the north were synchronous with periods of increasing rainfall, so-called pluvials, in the south. For instance, geological and pollen-analytical studies in the Staked Plain reveal a cyclic variation between steppe conditions, corresponding to interstadial or interglacial phases and times (corresponding to glaciations) when lakes developed in this now-barren area surrounded by park forests with pine and spruce as the predominant trees. And so, while the extinction of the big animals may be difficult to explain on climatic grounds, Lundelius thinks that the Recent withdrawal of the northern species—his second group—was due to a change in climate with increased seasonal contrasts, especially the hot, dry summers of central Texas today.

THE GIANTS OF THE PLAINS

Few of the Ice Age animals in America have caused so much disagreement among students as the bison. The number of species, the time of their first appearance, and their evolutionary relationships have been variously interpreted.

Take their first appearance. Studies of bison remains in the Pleistocene river terraces of Nebraska have suggested that they lived there as early as in Kansan times—well before the Rancholabrean. But other students contend that the complete absence of bison remains in well-documented Irvingtonian faunas contradicts this result, and that the stratigraphic position of the remains has been misinterpreted.

Take the number of species. Some students think there were as many as ten different species of bison in America during the Ice Age. At the other extreme, it has been suggested that all the American bison belonged to a single species evolving through time.

In the present book, I steer a middle course, but I want to point out that these differences in opinion exist and that the presentation here may be simplified, perhaps unduly so.

The Alaskan *Bison priscus,* an invader from Eurasia, was discussed in chapter 4. Apparently this animal migrated into America south of the ice, perhaps in the Yarmouthian interglacial or in an Illinoian interstadial, there to evolve into the mighty *Bison latifrons*—the greatest bison of all times. This superb beast was among the first Pleistocene animals of America to become known to science. It was described and named by R. Harlan back in 1825.

Bison latifrons had incredibly long horns. As in other bovids, the bison horn consists of a bony core encased in a horny sheath which forms the outer part and apex of the horn. The sheath does not normally fossilize (except in such special case as the Beringian mummies) and so we cannot say exactly how long the horns were in the giant bison. But the span, from tip to tip, of the bony horncores may reach a maximum of exactly seven feet (213 centimeters), which is more than three times the recorded maximum for a modern bison (2 feet, 2 inches; or 66 cm). The body size, however, was only about one-quarter larger than that of living bison.

Naturally, most giant bison were smaller than the prize specimen with the seven-foot horncore span. There is an extraordinary variation in the size of the horns, and in their shape too: some are much curved, others nearly straight. It is this variation which has led to so many different species being recognized. On the whole, however, it seems that there was a slow decline in horn size as time went by, and at the same time the horns tended to become more curved.

The giant bison was common and widespread in the Sangamonian, when it ranged from Mexico, Texas, Louisiana, and Florida in the south, to Idaho, Alberta and Saskatchewan in the north; and from California in the west to South Carolina in the east. But there are also Wisconsinan records, and the latest known giant bison comes from a locality called Rainbow Beach in Idaho, dating between 21,000 and 30,000 B.P. At that time, the giant bison had already been supplanted by a smaller species in most of the continent.

This smaller bison, usually called *Bison antiquus,* was still bigger and longer-horned than any bison now in existence. Its appearance at the end of the Sangamonian interglacial suggests that it represents a second *Bison* invasion from Alaska; it would in that case be a direct descendant of *priscus.* On the other hand, it could have evolved from some unknown marginal population of *latifrons* in which size had been reduced.

The history of *Bison antiquus* during and after the Wisconsinan is one of gradual size and horn reduction. Its average horn-core span (in male skulls—females were smaller) was 2 feet, 11 inches, which may be compared with a male average of 1 foot, 11 inches in modern plains bison. Bison of *antiquus* type survived well beyond the end of the Ice Age and grade imperceptibly into modern bison. This suggests that *antiquus* should be regarded as a subspecies of the living species *Bison bison.* As we have seen, the external characters of *priscus* differed considerably from those of *bison*, and it might be thought that the difference is too great to admit of a direct derivation; so this argument might favor an indirect descent, by way of *latifrons.*

In the Great Plains, vast herds of the *antiquus* bison race probably were a common sight during the Wisconsinan. With them may have mingled the typical camelid of the Rancholabrean, *Camelops hesternus* or "yesterday's camel." A bison herd killed by early man at Casper in Natrona County, Wyoming, 10,000 years B.P., included one individual of *Camelops. Camelops* was a member of the camelopine group, which was intermediate between llamas and true camels; in the flesh it resembled a dromedary with its single hump, but it was taller, with long legs and a long neck. These camels ranged to the Beringian refugium in the north, and to California, Mexico, and Texas in the south; but they were absent to the southeast, where llamas took over. The long-legged *Hemiauchenia* llamas, however, were present in the southern Great Plains.

The largest animals of the plains were the mammoths. In the Wisconsinan species, *Mammuthus jeffersonii,* the grinding teeth had reached a degree of specialization which makes it hard to tell them from those of the woolly mammoth; but the Jefferson mammoth was larger than its northern cousin and lacked the dorsal hump. It probably had shorter fur, too. Its remarkably twisted tusks, the points of which often crossed each other, have caused much speculation. Most probably they should be regarded as organs of display and ritualized combat. The fact that the tusks were lyrate in shape, with incurved tips, would ensure that normally no serious damage would be caused. They would thus play the same role in dominance-establishing bouts as the antlers of deer.

It can be objected that only male deer have antlers, whereas in mammoths, females as well as males had tusks. But we may note that female reindeer are antlered, and they use their antlers to defend their territory at the time of parturition.

If one of the tusks was broken off in an accident, the single-tusked mammoth could become a deadly adversary. This is, in fact, confirmed—and the theory of the tusks as organs of ritual combat supported—by a recent discovery which remains unpublished at the time of this writing.

Remains of the Jefferson mammoth have been discovered at hundreds of sites, and they range from northern and western Canada through the United States and into Mexico. Most are from regions where the environment seems to have been an open prairie, and it is thought that the mammoth fed mainly on grasses.

With a shoulder height of 11 feet or more, the Jefferson mammoth was one of the largest of its tribe. Yet there is good evidence, from kill sites and hunting camps, that early man preyed on it toward the end of the Wisconsinan, some 11–12,000 years ago. But man was not the only mammoth hunter. Recent discoveries have shown that there was at least one other creature which was able to prey on the strongest and mightiest of all the land animals in North America.

THE CAT AND THE MAMMOTH

The Staked Plain of Texas—the Llano Estacado of the Spaniards—rises gently to the southeast, up to the Edwards Plateau, the southernmost rampart of the Great Plains, between the Colorado and Pecos Rivers. This rocky, high-lying land is famous stock-raising country. But into its level strata of limestone many creeks and canyons have dug down, and they open the way to hidden caverns and fissures in which are secreted the remains of even weirder beasts than the famous Longhorns: the fauna of the Pleistocene.

The most remarkable of them all is the cave named for Albert Friesenhahn, on whose ranch it is located, not far from the village of Bulverde, in northern Bexar County about 21 miles north of San Antonio.

Precisely when Friesenhahn Cave was discovered, or by whom, is not known. The first publication dealing with it was penned in 1919 by E. H. Sellards, fresh from Florida, and in the following year O. P. Hay listed the fossil species found. But that early dig just scraped the surface. When excavations were resumed in 1949, with Sellards in charge, exciting discoveries began to accumulate. To Sellards, a diligent hunter of early man in America, those that pointed to man's presence were especially striking. For instance, a few flint flakes were found with the prehistoric animals. True, they could derive from flint nodules that had eroded out from the limestone roof or walls. But more mysterious is the shell of a freshwater clam from the same deposit. It was undoubtedly brought into the cave from the outside, and the nearest stream where it could have lived is several miles away.

Apart from these interesting possibilities, Friesenhahn Cave is unique in being the only site in the world where a mass occurrence of the great scimitar-toothed cat *Homotherium serum* has been found.

In contrast with the sabertooth *Smilodon,* of which thousands of bones have been unearthed, the scimitar cat has been cloaked in darkness and mystery. Tantalizing scraps of evidence have come from time to time all over the world—in North America, Europe, Asia, Africa: jaws, maybe a skull or two, mostly just one of the peculiarly shaped, scimitarlike fangs. A skeleton from the late Pliocene in France is the only reasonably complete find. There is enough to show that scimitar cats ranged far and wide. A few teeth show it to have been a contemporary of Neandertal Man in England. But little more can be said of it.

In the Texan cave were found skeletons of cubs and adults, and a number of bones. Clearly this was a lair of *Homotherium,* inhabited by scimitar cats for perhaps thousands of years: a retreat where they found shelter with their young, where they hid away when stricken by illness, and where they brought their prey. And so a flood of light was suddenly shed upon this mysterious animal and its habits.

The two sabertoothed cats of the Ranchola-brean. *Homotherium* the scimitar-tooth (above) lived also in the Old World. *Smilodon* (below) is only known from North and South America.

When these animals lived, the climate was perhaps not much colder than today, but probably more humid. Along with the great grazing animals, like bison and mammoth, were found remains of browsers such as the American mastodon and the peccary *Mylohyus,* of which a complete skeleton was excavated quite close to that of a big scimitar cat. A relatively high humidity is also suggested by the fact that much of the sediment accumulated under water, so the water table was higher than today; there must have been intermittent ponds on the cave floor. The abundance of turtles (at least 354 individuals!) is intriguing. Possibly they hibernated in the cave, but they may also have been attracted by the ponds.

The first to recognize that North America had harbored a second great saber-toothed cat besides *Smilodon* was Edward D. Cope, back in 1893; his material, a few teeth and foot bones, came from western Oklahoma. He gave the cat a generic name of its own, but later studies have shown that it was so closely related to a European form called *Homotherium* that it has to be considered the same genus. As *Homotherium* was named as early as 1890 by the Italian Emilio Fabrini, *Homotherium serum* it is.

The name means "late similar beast," but although *Homotherium* at first sight shows some resemblance to *Smilodon*, the differences are many and striking. Looking first at the great fangs, those of the scimitar cat are shorter, more curved, and extremely thin, with beautifully serrated edges somewhat like a very fine-toothed saw-blade; and such edges are also found on the other teeth. Clearly these teeth form a very finely gauged apparatus for the cutting and slicing of red meat. On the other hand, they could not possibly work on bone without being blunted. The *Smilodon* teeth are stouter and, except for the sabers, lack serration.

In its body build, too, the scimitar cat was different. Most striking is the extreme elongation of its front legs, especially the forearm and hand, while the hind limb is short. The animal evidently stood very high in front, with a sloping back and probably a raised head. *Smilodon*, on the other hand, is usually pictured in a low stance, with its head close to the ground.

What did *Homotherium* do for a living? The debris of its prey gives some very revealing hints. For how else can we account for the hundreds of mammoth teeth found in the cave deposits? The mammoth teeth are *milk* teeth. (There are a few milk teeth of mastodons too, but the great majority are mammoth.) They come from very young individuals.

There are stories about "elephant graveyards" where elephants retire to die. Students are somewhat skeptical about this, and in any case the bulk of the individuals would be old adults.

We can hardly expect mammoth calves to troop into the cave to die. Most probably they were killed elsewhere, and brought into the cave as prey. And in that case there is is no animal other than the scimitar cat who could have done it. Friesenhahn Cave is the the sole known "regular" lair of *Homotherium,* and it is also the sole known place where mammoth calves occur *en masse.* This cannot be a coincidence.

Certainly, the scimitar cat was well equipped to deal with a mammoth carcass. Lots of red meat there; no need to blunt those precious teeth on bones.

But here we come up against a facer. How did the scimitar cat manage to kill mammoth calves without being killed itself by the enraged parents? Modern elephants are highly social animals; they live in herds and take very good care of their young.

Scimitar cat Homotherium, attacking baby mammoth.

Perhaps we should not apply elephant standards to mammoths. They are not as closely related as we used to think. In fact the mammoth lineage has been separate from the two elephant ones for some five million years. Probably their habits differed in various ways. The mammoths, for instance, may have been solitary or formed small family groups rather than herds. The prolonged success of *Homotherium* as a mammoth hunter certainly indicates that mammoth calves were an easier prey than living young elephants.

My bet is that the scimitar cats hunted in pairs. (To drag a dead young mammoth into the cave would probably need the combined efforts of two cats!) If one of the pair succeeded in tricking the adult mammoth parent (or parents) away, the other could kill the calf from an ambush. That could be the meaning of those slender, deadly scimitars. A quick slashing stab into the neck, for instance, followed by a rapid retreat, would leave the prey to bleed to death. (This would also account for the high stance of the head and forequarters.) The victim could then be left until the adult mammoths vacated the scene.

Social hunting among cats is rare. The lion, as we have seen, is almost the only living example. Yet there is evidence that cheetahs may occasionally hunt in pairs. Also, recent studies on the rare snow leopard of central Asia have suggested that pair-bonding may occur. Both of these great cats, incidentally, are comparatively brainy. *Homotherium*, too, had a fairly large brain.

All of this is frankly speculative, but it is supported by the meager evidence given by the dead bones. So fascinating is the possible truth lurking behind the discoveries in Friesenhahn Cave that one is unavoidably tempted to strive for a glimpse of it.

Sometime, somewhere, human beings must have seen scimitar cats on the hunt, but we can only guess at what they saw. Scimitar cats ranged widely in the Rancholabrean. They were in Alaska at the same time as early man. Their remains, mostly scanty, range from California to Florida. Perhaps they were too smart to become common fossils.

The cats that left their bones in the cave fall neatly into two groups. On the one hand, there are cubs and half-grown individuals. On the other, there are old animals with excessively worn teeth. But there are hardly any remains at all of adults in their prime.

This is the typical pattern of natural mortality. The immature ones are probably cubs that had lost their parents, and were left alone in the cave to starve to death. The occasional half-grown or nearly adult individual may have starved from inexperience. The very old ones, incapacitated by the ruin of their once splendid armament, may have sought the cave only to be overcome by death. To all, probably, the cave meant familiarity, shelter, a place to rest in safety, and perhaps to recuperate.

No doubt animals in their prime died, too, but that happened elsewhere. We may imagine that they sometimes had the worst of it in a bout with the mammoths. Up against the most powerful animal of its world, the scimitar cat led a dangerous life.

THE RESURRECTED PIG

"Living fossils": the term is usually applied to animals and plants which have remained virtually unaltered for very long stretches of time—organisms which, so to speak, evolution has passed by. A famous example is the little marine lampshell *Lingula*, which looks essentially as it did half a billion years ago. (The same, incidentally, is true for some other lampshells, notably the genus *Crania*, which for some reason has never attained the same fame.) This does not really mean that evolution has come to a perfect stop; the forms of the distant past, although very like the living, are not identical with them. What it evidently does mean is that these organisms have hit upon a very enduring mode of life, in a stable environment, and once adapted to it have no need of radical change.

But the term could be given a different meaning: an animal or plant only known from fossil remains, thought to be extinct, and then discovered alive. This is not as unusual as one might believe. The beautiful, ribbed bivalve *Trigonia*, so common in the Jurassic and Cretaceous seas of the northern continents, was long thought extinct, but in 1802, a French explorer found a living descendant in Australian waters. The fossil lungfishes, *Ceratodus*, in America and Eurasia, were also found to have living descendants in Australia. The peculiar goblin shark or elfin shark—a weird-looking 12-foot creature with a spearlike nose and a great tailfin—has the same history: originally known in the fossil state, it was then found living, first in the Pacific, then in the Indian and Atlantic Oceans.

There are mammals, too, in this category. The okapi of Africa, discovered around the turn of the century, belongs to a stock of giraffes that was known from fossil

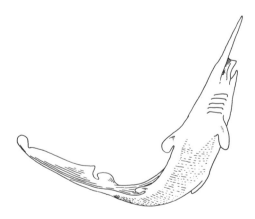

The goblin shark—a living fossil.

remains long before that. In 1966, Australian zoologists were astonished to find a living member of the small marsupial *Burramys*, till then only known as a Pleistocene fossil. Examples from South America are the "rat" *Blarinomys* and the bush dog *Speothos*. But the case of the flat-headed peccary is even more spectacular.

Its story starts in 1804 or 1805, with the find of a skull in a Kentucky cave. The first to study it was no less a person than Caspar Wistar, the illustrious Philadelphia physician and scientist who was later to succeed Thomas Jefferson as the president of the American Philosophical Society. His name is commemorated in the plant genus *Wisteria* (related to the peas)—and, of course, in the Wistar Institute founded by his grandson. In 1806, he identified the skull as belonging to a peccary.

Peccaries, or "American pigs," are in fact related to Old World pigs, but have a long separate history. Originally, they were present in the Old World too, but there they died out long before the Pleistocene, perhaps as a result of competition with "true" pigs. You can tell a peccary from a pig by its tusks: they curve outward in pigs, but not in peccaries.

The animal remained nameless until 1848, when it was dubbed *Platygonus compressus* ("broad-jointed, compressed") by John L. Le Conte. The leading American entomologist of his time, Le Conte was also interested in a variety of subjects ranging from geology to ethnology.

Since then, finds of flat-headed peccary have accumulated from almost every corner of the United States, and it appears to have been one of the common animals of the Wisconsinan ice age. It is frequently found in open-air deposits, but the real mass occurrences come from caves, where *Platygonus* may make up 90 percent or more of the fauna. Bat Cave in Pulaski County, Missouri, contained the remains of at least 98 peccaries, and Zoo Cave in Taney County (also in Missouri) contained 81, of which half were young animals with milk teeth.

The genus dates back to early Blancan times. Soon, flat-headed peccaries invaded South America, and gradually the lineages in the two continents diverged from each other. The North American forms became very long-legged, and are regarded as fast-running animals of the open plains. Such extreme specialization was not attained in South America, and that lineage is now regarded as a distinct genus, *Catagonus*, descended from early *Platygonus*.

Catagonus is common enough in the Pleistocene of South America. But there is one site in the province of Santiago del Estero, Argentina, where one species *(Catagonus wagneri)* was found in association with pre-Columbian archaeological material. So *Catagonus* survived the end of the Pleistocene in that area. Still, to find it alive and kicking was a great surprise.

The discovery was made by Ralph M. Wetzel of the University of Connecticut and his associates in 1974—170 years after the initial discovery of fossil flat-headed peccary. This was in the Chaco —a scrub-thorn and grass region—of western Paraguay, but it is now known to live in other parts of the Chaco too, in northern Argentina and southeastern Bolivia. Called the tagua (or pagua) by the Chaco Indians, it is actually one of their favorite game animals, and even seems to be on the increase, probably due to the killing-off of its chief predator, the jaguar.

It seems remarkable that the tagua escaped detection for so long, in spite of the efforts of diligent traveling naturalists. Perhaps field mammalogists have a tendency to be on the outlook for small mammals, rather than large; no one will readily believe that big unknown beasts are still stalking about. Now it may well be suspected that some peccary specimens in old collections (skins or skulls), will turn out on reexamination to be tagua, instead of the two long-recognized living species, the white-lipped and the collared peccary. Both are in the genus *Tayassu*, and only distantly related to the flatheads.

The tagua, then, although not the same animal as *Platygonus* of the north, is closely enough related to it to give a fair idea of what these animals may have looked like in life.

The *Tayassu* peccaries are nocturnal in their habits, and adapted to a forest environment. They have relatively small heads with a short snout. Wetzel sees them as advanced offshoots from the *Platygonus-Catagonus* stock. The tagua, on the other hand, retains more diurnal habits, is a browser rather than omnivorous (contrary to *Tayassu*), and is somewhat larger in size. Its long muzzle is turned downward to a greater extent than that of *Tayassu*, which permits it to scan the horizon when browsing; the same feature is found in *Platygonus*, and is clearly useful to an animal of the open spaces. An interesting detail is the nasal cavity, which is greatly developed in *Platygonus* and *Catagonus*, and probably functions as a dust trap, keeping the sense of smell intact in a windy, dusty environment.

The most important difference between the South and North American flatheads is that the latter, especially the Rancholabrean *Platygonus compressus*, are distinctly longer-limbed, and probably more fleet of foot than their southern cousins. The adaptation to movement over hard ground has led to the complete loss of the dew-claws in *Platygonus*, while *Catagonus* retains them on the front limb. *Tayassu* has the full complement of dew-claws, which gives it a spreading foot well-suited to the softer ground in the forest.

The tagua in the flesh is a large-headed animal. The Indians have a saying that if you cut off its head, you are left with only half the animal. The predominant color is brownish-grey, but there is a white collar band extending back to the shoulders. The hairs are much longer than in *Tayassu*, giving the animal a shaggy appearance.

In North America, the flat-headed peccary *Platygonus compressus* ranged north to the vicinity of the Wisconsinan ice-margin, and south to Mexico and Florida; and it has been found from California in the west to New York in the east. The animal stood two-and-a-half feet at the shoulder and compared in size with a modern wild hog, but had longer legs and—as already noted—lacked the side toes. It was adapted to fast movement on hard ground. Enemies that could not be shaken off were probably attacked with the sharp canine teeth, which certainly could give a very nasty bite. But there were other dangers beside predators.

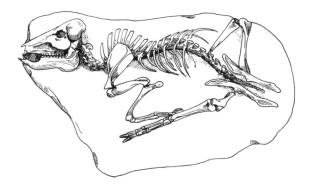

This flat-headed peccary was one of five discovered at Hickman, Kentucky. Lying as it succumbed in a dust storm more than 34,000 years ago, it had turned its back to the wind in an effort to avoid the stinging, choking sand.

The peccaries roved the plains in small herds or "sounders." At Hickman in western Kentucky, a group of five peccaries was found; they were lying as they had perished, in a row extending about ten feet, and buried deep in a windblown sediment. They died more than 34,000 years ago, at a time when the ice-margin was about 200 miles away. The area was grass-covered (traces of the roots can still be seen), but there were probably some spinneys around, for the cracks in the ancient land surface have preserved numerous seeds of the forest tree *Celtis occidentalis.* The animals had been heading east up a slight incline, probably coming from the Missouri River of that time, only a few miles away, when they were suddenly overtaken by a violent, and probably very cold, northwesterly storm. To protect themselves against the choking dust, all the animals had turned their heads to the southeast. But it was to no avail. They perished and were covered by dust, piling up to ten feet over the centuries and millennia.

Other peccary groups, probably killed in the same way, have been discovered in Colorado, Michigan, Kansas, and Ohio. Dust storms were evidently a common hazard in the Rancholabrean Great Plains.

There is no evidence that man hunted the peccary. The most recent date for *Platygonus* comes from Mosherville, Pennsylvania, and is 11,900 B.P. On this basis, it seems that the flat-headed peccary of North America became extinct somewhat earlier than the majority of the Pleistocene forms.

HORSES, ANTELOPES, AND PREDATORS

For more than a century, it has been known that wild horses existed in North America in the Pleistocene. In 1858, Joseph Leidy described two species, and since then an extraordinary number of species names have been bestowed on Pleistocene horses. The result is a puzzle even more intricate than the bison one. The systematist tends to get lost in a jungle of names, many of them based on a single tooth of quite uncertain age.

Fortunately, recent efforts, especially by Morris F. Skinner of the American Museum of Natural History and Walter W. Dalquest of Midwestern University, are creating order out of what was chaos.

After the extinction of the Blancan gazelle-horses, zebras, and true asses, three major groups of American horses took over. One of them consists of long-legged half-asses whose close relatives still live in Asia—onagers, kulans, and kiangs. The other two groups are also related to asses but are somewhat more distant from the living Old World forms. They differ in details of the cusp patterns of the cheek teeth; some were long-legged (though never to the degree seen in the onagers), others had short and stocky limbs. Some of them grew to the size of modern draft horses, others were no larger than living donkeys. Although much remains to be done before their history is properly understood, it is evident that there were many different species and that a variety of modes were explored.

There is some regional differentiation. Among the onagers, for instance, the extremely stilt-legged forms were only found in the south, while Beringia was inhabited by the still living Asiatic species *Equus hemionus*. The short-legged, very small *Equus conversidens*, on the other hand, ranged from Mexico to Alberta, and seems to have been one of the commonest and most widespread species. A large species, the western horse or *Equus occidentalis*, is mainly found along the Pacific coast, while the great *Equus giganteus* and some related forms were predominant in the interior.

At least some of the species survived to the end of the Wisconsinan, or even later. At one locality in Alberta, *Equus conversidens* may still have been in existence about 8,000 B.P., while larger horses of the *Equus occidentalis* type have been found in Idaho in a cave deposit dated at 10,370 B.P. These horses were hunted by man.

The variety of antelopes in the Great Plains and adjoining areas is remarkable, too, although these are restricted to the southern part. They were plains forms and, unlike the horses, did not range into Florida in the Rancholabrean. All were related to the living pronghorn (not to the Old World antelopes), but most were somewhat larger and the horns were often bifurcated, so that the animals were four-horned. Several genera and species can be distinguished.

The fastest predator of the Great Plains was the American cheetah, *Miracinonyx trumani*, a descendant of the Blancan form and a sibling of the living Old World cheetah. Lions, sabertooths, pumas, wolves, and dire wolves were other large carnivores which also took their toll of the plant-eaters.

There was also a bear of the tremarctine stock, related to the Florida cave bear but very different from it. This was the great short-faced bear, *Arctodus simus*. It was larger than any bear living today (or in its own time): even the great Alaskan brown bears would look small beside it. In contrast with the cave bear, *Arctodus* was remarkably long-legged and slim of build, and must have been unusually fleet of foot for a bear. Yet the *Arctodus* bears were so large that some of them probably would have

Artist's impression of the heads of two extinct pronghorn antelopes—*Capromeryx* (top) and *Tetrameryx*.

tipped the scale at 1,500 pounds. With its broad, high-vaulted head, *Arctodus* differed from the narrow-skulled, long-nosed ursine bears; the shape of the skull is actually reminiscent of that in a great cat. The back teeth were only moderately enlarged for a bear, and it is thought that *Arctodus* was mainly a flesh-eater. It may well have been the top predator of its world.

The history of *Arctodus* began in the Irvingtonian, and in the Rancholabrean it was present over most of the continent, including Beringia. It was long thought to be absent in Florida, but recently a few finds have revealed its presence; it was probably rare, however, compared with the cave bear and black bear.

The demise of *Arctodus* toward the end of the Wisconsinan has been coupled with the entrance of the grizzly at the same time. The two may have been in competition. However, in Beringia *Arctodus* held its own, probably for a hundred thousand years or more, in spite of the presence of the grizzly.

The marvelous wildlife of the Great Plains in the Wisconsinan has no equal today. Even that of the African savannas is eclipsed. Now they are all gone, even the bison herds that were so thick only a century ago. We can form only an incomplete image of the majesty and richness of this animal world which vanished so long ago from the living memory of man.

THE GREAT BASIN

Crossing the Rocky Mountains and the Colorado Plateau, we come into the Great Basin, forecourt to the Pacific coast. In this area, many ancient lake shores have been found in direct connection with glacial outwash, and they show that the lakes were much larger than now. The Pleistocene lake district has the shape of a triangle, the corners of which were formed by three great lakes.

One of them is Utah's Great Salt Lake, whose Pleistocene predecessor, Lake Bonneville, rose up to 1,000 feet above the present level and covered an area many times as wide. At its highest level, Lake Bonneville debouched to the north into the Snake River, scouring out a 300-foot canyon. The weight of this body of water was sufficient to depress the crust of the earth; the unequal rebound has now caused the old shorelines to become distinctly warped. The sequence of lake-bottom sediment and shorelines make it possible to reconstruct the history of Lake Bonneville, and it shows that the lake waxed and waned in synchrony with the land ice.

Farther west, in northwestern Nevada, developed the intricate system of Lake Lahontan, of which present-day Pyramid Lake is a small remnant. Other pluvial lakes were in Oregon and northern California.

South of these inland seas, lakes dotted the interior of the Great Basin, even areas which are now among the driest and most forbidding in the world. Even in the grim Death Valley there then sprawled Lake Manly, a hundred miles from end to end. Its name honors William Manly, who saved a covered wagon party there in 1850.

A southern outpost of the lake district, Lake Cahuilla, formed a third inland sea, of which the Salton Sea is today's remnant. The Colorado River debouched here in the Salton Trough, originally an extension of the Gulf of California, which was then isolated by a delta barrier built up by the river. As the meandering river changed its bed, the trough filled up or dried out.

Many remains of the Pleistocene life of this area have come down to us. The fauna was essentially the same as that of the southern Great Plains. However, there is one remarkable creature whose life zone was restricted to the Great Basin and adjoining areas. This was the shasta sloth, *Nothrotheriops shastensis*.

In the Irvingtonian it ranged east to the Texas Panhandle, but Rancholabrean records are limited to an area extending from Alberta in the north to Mexico in the south, and from California in the west to New Mexico in the east. Many finds come from arid caves. Two are especially worthy of note.

A skeleton of the shasta sloth was found in Aden Crater, Doña Ana County, New

Mexico. This is an extinct volcanic crater and the sloth was discovered about 100 feet below the surface. In the dry climate, some of the soft parts of the body were preserved, and tendons, ligaments, claws, and patches of the skin with hair still adhere to the bones. The specimen is now in the collections of the Yale Museum.

Like the hermit sloth, the shasta sloth belonged to the Megatheriidae. It was, however, the smallest of the North American ground sloths (not counting a number of dwarfed megalonychids on the Caribbean islands). The Aden Crater specimen, an immature animal, measured about five feet in length, excluding the tail. The weight of a fully grown animal is estimated at between 300 and 400 pounds. The shasta sloth was much lighter in build than the others, in particular the front legs were long and slender—almost spidery, says Professor Scott. It walked on the knuckles of the fingers. It had a small head and a rather long neck.

The hair covering is remarkable. The long, coarse hairs are yellowish, but there is evidence that they were infested with algae, as in the living tree sloths, giving them a green tinge.

Enormous amounts of *Nothrotheriops* dung have been preserved in some dry caves, the most famous being Rampart Cave. Discovered in 1936, this cave is situated on a cliff at the south side of the Colorado River just below the Grand Canyon, and about 500 feet above the present river level. The entire floor was found to be composed of sloth dung, layer upon layer, and hundreds of bones were entombed in this extraordinary sediment. Besides the sloths, remains of many other mammals were found, among them jackrabbits, woodrats, ground squirrels, desert mountain sheep, puma, and bobcat, all still occurring in the area. Extinct animals include the horse and the little mountain goat *Oreamnos harringtoni,* named for Mark Harrington, one of the explorers of the cave. Condors and owls had inhabited the cave, too.

Harrington, an anthropologist, tells of the "unmistakable smell of ground sloth" pervading the cave. The dung is made up of typical desert plants such as cacti, yucca, agave, and orache. Radiocarbon analysis showed the top layer of the dung to be 7,900 years old, while the lower layers gave progressively higher ages down to 38,300 at the 54-inch level. The cave had thus been inhabited by shasta sloths throughout the later part of the Wisconsinan. It has been suggested that the sloths used such dry caves as winter dens, seeking higher ground in the summer.

I have to conclude on a sad note. Although the entrance to the cave had been barred, in 1976 vandals forced their way into it and set fire to the deposit. The fire destroyed much of its irreplaceable content. Even fossils are not exempt from the vagaries of human folly.

VII

Tar Pit Country

RANCHO LA BREA

IN AUGUST 1769, Gaspar de Portola, riding with an exploring party in what is now
the city of Los Angeles, was surprised to observe to his right an area with lakes
and swamps of black bitumen. He was intrigued by this substance, which appeared
to flow out from underneath the earth, and even wondered if it could have something
to do with the earthquakes so common in the region. But it remained for his successor, another explorer called José Longinos Martinez, to discover bones in the tar in
1792. He describes

> a great lake of pitch, with many pools in which bubbles or blisters are continually
> forming and exploding. . . . In hot weather animals have been seen to sink in it
> and when they tried to escape they were unable to do so, because their feet were
> stuck, and the lake swallowed them. After many years their bones have come up
> through the holes, as if petrified. I have brought away several specimens.

Fifty years later, the French explorer Duflot de Mofres noted that animals approached
the water pools to drink, in spite of the mineral taste of the water, and so ran the risk
of getting stuck in the tar underneath.

These were the first descriptions of what was to become famous as Rancho La
Brea, Major Henry Hancock's ranch in the late nineteenth century. It is now known
as Hancock Park and has long since been engulfed by the growing city. But though
the site is now a well-kept park, the pitch still forces its way up through lawn or

footpath, and the bubbles rise all over the reflecting pool. The tar seep of Rancho La Brea still lives. Others in the region may still be active as traps, like Carpinteria in Santa Barbara; or they may be dead, as McKittrick in the San Joaquin Valley apparently is. Dead or living, they contain incredible masses of bones in their black interior, and to this day excavations are going on to uncover a wealth of information about the life of the past in California.

The California tar seeps are remarkable, but not unique. In Peru, the Talara asphalt deposits have yielded great numbers of Pleistocene animal bones, many of them belonging to the same species as those of Rancho La Brea. In the Old World, a fossiliferous bitumen deposit is known at Binagady in the Caucasus. But no other place is even remotely as prolific as Hancock's ranch, from which thousands upon thousands of bones have been collected. So important is the faunal record extracted from these deposits that the late Pleistocene Land Mammal Age of North America, spanning the last half million years or so, is called the Rancholabrean Age.

This is oil country, and the oil comes from marine deposits that underlie the Pleistocene beds. These sea-bottom strata, dating from the Tertiary period, have been folded and fractured by the same forces that still cause earthquakes: the connection imagined by de Portola is there, although not quite in the way he thought. These are also the forces that cause the seaward part of California to slip gradually to the northwest along the San Andreas fault.

The oil trapped in the strata forces its way upward through cracks and fissures. When it reaches the surface, it loses its lighter components through evaporation; the residue, flooding the areas around the vents, is the pitch or brea. Especially if covered by a sheet of water, it may now become a deadly trap to unwary visitors.

Animals come to drink, get caught, and struggle to break away. The struggle promptly catches the interest of every flesh-eating animal in the neighborhood. The fossil-collecting trap is baited. Here comes the great dire wolf, the hyena of the New World, the hungriest of them all; and he gets caught, and perishes. Here comes the monster sabertooth, and he gets caught, too. Soaring far away, condors and vultures sense the commotion and swoop in; many are added to the bag. Insatiable, the tar pool pulls its victims down, one after the other, strong or feeble, in a chain-reaction which ends only when there is no meat-eater left nearby to succumb to the lure.

Years pass. The slow eruptions of newly risen oil stir the viscous mass, causing the mired bodies to rip apart, mixing the remains of predator and prey. New vents open and old ones dry out. The tar hardens and gradually becomes buried under the dust and silt of millennia. The deposit has come to a rest. It remains for us to find it, and the record of past life within it.

This is the classic picture of Rancho La Brea. Recent studies have modified it to some extent: the disarticulation and abrasion of the bones is now thought to be due to the action of running water rather than to movements within the tar. The fossils do not occur in the open asphalt lakes but in tar-impregnated, water-laid sediments. Also, it is evident that animals were entrapped less frequently than originally thought.

The mass accumulations result from the entrapment of a few individuals per annum during a very long time, perhaps about 30,000 years, in asphaltic quicksand for instance. Still, the great preponderance of flesh-eating animals shows that the effect was selective.

From radiometric dates of the bones and pieces of wood we know that the animals found lived here during the last, or Wisconsinan Glaciation. The various pits and "pockets" evidently were active at different times. One, for instance, gives dates ranging from 40,000 to 25,000 years B.P. Another dates from 19,300 to 12,650. With the fluctuations of sea level and climate, the scene changed repeatedly, as witnessed by the plant remains, which suggest an oscillation between a warm, dry, inland vegetation and a cool, moist, coastal flora culminating in redwood forests. But on the whole, plains animals predominate greatly in numbers over forest-dwelling species.

DIRE WOLVES AND SABERTOOTHS

Although more than 40 species of mammals and 120 of birds have been found in the bitumen of Rancho La Brea, two mammals greatly surpass the others in abundance: the dire wolf and the sabertooth. Some years ago Leslie Marcus made a census of the specimens from the various pits, estimating the minimum number of individuals. His final count for dire wolf was 1,646, for sabertooth 1,029. Among mammals, the coyote came in a poor third with only about 240 individuals. The bird census made by Hildegrade Howard shows that a number of bird species, headed by the golden eagle, beat the coyote. Almost all of these birds and mammals were meat-eaters: an extinct turkey is the only exception.

The dire wolf was first discovered and named *Canis dirus* by the great pioneer of American paleontology, Joseph Leidy, more than a hundred years ago. His material came from Evansville, Indiana, but the same great wolflike creature has since been found over most of the continent, from Idaho in the north to Mexico and even Peru in the south, and from California in the west to Florida and Pennsylvania in the east. It is lacking only in the far north, and evidently was not equipped for a life on the tundra.

We may picture this animal as a largish wolf, rather heavier and shorter-legged than a timber wolf, and with a strikingly big head, armed with immensely powerful teeth and jaws. It clearly was not quite as fast and indefatigable a hunter as the living wolf, and it probably preyed on large and comparatively slow animals. Its life habits may have been somewhat like those of the spotted hyena of Africa, which is known to be a fierce and fearless hunter as well as a remarkable eater, dealing competently with the tough skin and heavy bones that no other animal can chew.

It is possible that the dire wolf hunted in packs. Helmut Hemmer has shown that, contrary to earlier opinions, its brain size surpassed that of the wolf and the Cape hunting dog—both highly skilled social hunters—which may suggest an even more highly developed social organization. Yet, in spite of its apparently high intelligence, the dire wolf became extinct. The youngest date for a dire wolf from Rancho La Brea is 9,860 ± 550 years. (There is a slightly more recent date from Brynjulfson Caves, Missouri: 9,440 ± 550, but the difference is overshadowed by the standard errors.) Perhaps the end of the species was due to disappearance of its prey. There is indeed a tendency to associate the dire wolf with the big extinct bison race, *Bison bison antiquus.* Be that as it may, we must lament the fact that this superbly endowed great dog is no longer with us.

Where did it come from? We do not know for sure; the history of wolflike dogs in North America is still full of holes. As we have seen, information on coyotes is better; we can trace its history back to its beginnings with the "rabbit-hunting" dog of Cita Canyon and Mt. Blanco. The great Rancho La Brea coyote, an animal almost the size of a wolf, is a link in this evolutionary chain, and his smaller descendants still stalk their prey in the wilds of California. Was he already as partial to juniper berries as the coyotes of today? Or is the taste a late acquisition, like that of man for gin?

What happened to them when the sabertooth arrived on the scene? This is *Smilodon fatalis,* the late descendant of the *Megantereon* dirktooth of Blancan times. Did the dire wolves retreat in shame, crouching with glowing eyes as the monster cat came up to the carcass of a bison or mammal that happened to bait the trap? Or did they stand their ground, whipping up their excitement to the point where they assaulted the cat and drove it away? We can only guess at the dramas enacted around the tar pools.

The sabertooth is perhaps the most fantastic animal in the mired zoo of Rancho La Brea. Not only its outlandish appearance, but also the sheer numbers of its remains tend to boggle the mind. Imagine a thousand smilodonts coming to life around you! The size of an African lion, it differed by its shorter-legged and heavier build, and quite particularly by the sabers formed by its upper canine teeth, glisteningly visible even when the jaws were closed. They then protruded well beneath the chin. In this the smilodonts differed from their predecessors the dirktooths, for as we have seen their canines were sheathed by the chin.

Every feature here shows an animal of measured and stately movement, of great strength and sturdiness. This creature did not hunt by speed; nor, we may imagine, did it go out of anybody's way. Attacking with its powerful paws and its sabers, it could maim a large prey, rip up its neck or belly, and feed on the soft entrails and red meat, leaving the bones to blunter-toothed rivals.

The first to find a *Smilodon,* and to name the genus, was the Danish explorer Peter Vilhelm Lund, more than a century ago. His collection, still to be seen in the museum in Copenhagen, came from Brazil. The smilodont from Argentina and Brazil is a distinct species, even larger and heavier than *Smilodon fatalis;* it is almost a caricature, with lengthened sabers and shortened, immensely broad feet. The West coast species in South America seems to be the same as the North American.

In North America, the sabertooth ranged far and wide, and its range is very similar to that of the dire wolf; like that animal, too, it has not been found in the far north. Elsewhere, the ubiquity of *Smilodon* contrasts with the elusiveness of his distant cousin the scimitar cat, *Homotherium.* Was the smilodont so much more common than the homothere, or is it just that he was the more prone to become a fossil? Actually, what do you have to do to become a fossil? Well, generally speaking, you have to be a dope, that's what you have to be: you have to get caught in a tar trap, or drowned in a river, or lost in a cave. Perhaps the sabertooth was more of a dope that the scimitar cat.

There might be something to it. Hemmer shows that the brain size—or more properly the "degree of cephalization"—of *Smilodon* does not reach the level of a highly social cat like the lion but is in the range of nonsocial living species like pumas, leopards, and jaguars. A supercarnivore in terms of armament, and a highly aggressive creature as suggested by the large number of bones injured during life, it may have met its match in the more highly organized dire wolf packs.

Dope or not, the smilodont has a skull the stark functionalism of which inspires wonder. Everything is built around those two sabers, those enormous, thin, recurved blades of living ivory, with their serrated edges. The whole architecture of the skull and the neck, and indeed the entire animal, is geared to serve the supreme moment of stabbing.

The lower jaw had to be opened, swung down into a 95-degree gape, so as to clear the points of the sabers. Yet the curve of the sabers is such that the stab, or bite, could not penetrate very deep in the body of the prey. On the contrary, the sabers went in just under the skin and parallel to it. By ripping backward, the smilodont was able to slit up the entire belly or neck of its victim.

That is why the sabers became so big, and how the cat used them. In the early twentieth century, the great American paleontologist William Diller Matthew argued that the sabers were useful to their possessors and so would have been favored by natural selection. But at that time, such a Darwinian explanation of evolution was unpopular, and Bergsonian and other vitalistic and finalistic ideas were rife: evolution was thought to result from some inner force, directed toward a predetermined goal. So Matthew's authoritative discussion was promptly forgotten, and *Smilodon* started upon a peculiar career in philosophical, psychological, ethical, and other unexpected connections, as a warning example of evolutionary derailment.

Thus arose the *Smilodon* myth. According to it, the sabers grew so big that the poor animal could not open its mouth (or, alternatively, could not close it) and, so the myth went, it had been driven into this desperate situation by a mystical evolutionary force running its own inexorable course without regard for the fate of the vehicles in which it was embodied.

Biologists do not fall for this kind of balderdash any more. But the myth still pops up in unexpected places. Our predilection for romance being what it is, we may expect to see myth keeping just ahead of denial, vitalistic *Smilodon* being chased all over the

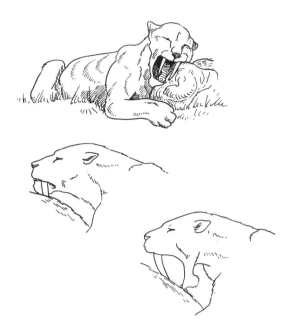

Sabertooth *Smilodon* after the kill: how did it cut up its prey? Margaret Lambert Newman suggests three possibilities: (1) (top) by using its cheek teeth; (2) (middle) using protruding points of sabers with mouth closed; (3) (bottom) using sabers with mouth open. Mode 2 seems mechanically very efficient—but *Smilodon* was the only sabertooth which could have used it, for in other forms (such as *Homotherium*) the tusks do not protrude beneath the chin.

world by rationalistic *Smilodon*, for years to come, somewhat like the two versions of the priest-detective in Chesterton's *Scandal of Father Brown*. When *Smilodon* finally became extinct, the cause was not evolutionary derailment. It died out because the large mammals on which it preyed had vanished.

Many other carnivores occur in the brea deposits, but they are rare in comparison, although the 76 lions counted by Marcus form an imposing array in themselves. We have already met the lion and the great short-faced bear. Puma and bobcat are also found, but the jaguar, so common in the forests of Florida, is extremely rare.

HORSES AND ANTELOPES

Among the plant-eating mammals, too, species of the open country predominate greatly, while forest animals like deer and tapirs are rare. The most common big grazer is the bison (159 individuals in Marcus's census), followed by the horse (130). The latter is the widespread Western horse of the Pleistocene, *Equus occidentalis*. With a height at the withers of about 14½ hands on average (4 feet 10 inches, or 150 cm. to those of us who are not horse fanciers) it was about the size of a modern Arabian, but certainly a lot heavier in build. Walter Dalquest feels that, in common with most other Pleistocene horses of North America, it was more closely related to today's asses than to true horses.

The camel of yesterday, *Camelops hesternus*, is also present (36 individuals), and so is a remarkable little antelope related to the living pronghorn. The living species,

Antilocapra americana, is also present at Rancho La Brea, but the great majority of the antelope remains (34) belong to the extinct, graceful *Capromeryx minor,* which was about one-third smaller. Although the horn of the living antelope is forked, its bony core is not; but in *Capromeryx* the horn-core is deeply forked into two distinct prongs.

The short, light body and long slender limbs of the diminutive brea antelope attest to its great speed. A herd of *Capromeryx* pronghorns, silently skimming the prairie in rapid flight, must have been a sight of extraordinary beauty. It seems a shame that we now have to conjure up such lost splendor from dry bones. Why should this delicate creature become extinct? Surely not from the depredations of the brea carnivores, however formidable they may have been. None of them could have been fast enough to run down a *Capromeryx* on the hoof.

THE BEAST OF THE TONGUE

Next to the bison and horse, the great ground sloth *Glossotherium harlani* is the most common plant-eater found in the tar (76 were found). Glossotheres (Greek for "tongued beasts") have been known to science for more than a century. The name was given by Sir Richard Owen, Darwin's contemporary and putative enemy, in 1840. The trivial name commemorates one of the pioneers of North American paleontology, R. Harlan.

Harlan's ground sloth was a grassland animal, ranging over most of the continent and up to the state of Washington in the north; a related species is in South America. With its impressive length of 11 feet, it was a medium-sized animal as ground sloths go. It belongs to a family of ground sloths, the mylodonts, which was mainly distributed in South America, its land of origin; but it had a foothold in North America since Blancan times.

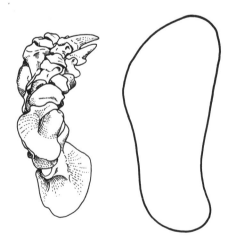

Tracks of ground sloth *Glossotherium harlani* are preserved near Carson City, Nevada. As might be expected, skeleton of sloth foot fits well into track.

The glossothere was a very hefty creature with great forelimbs, massive hind legs, and a stout tail. It walked in an ungainly manner on the outside of its feet. Fossil tracks of this animal have been found in Nevada, near Carson City, oddly enough in what is now the yard of the Nevada State Prison. Individual tracks reach a length of up to 19 inches.

Small-brained and small-eyed, but with an excellent sense of smell, this animal found protection against lions and sabertooths not only in its large size and powerful, clawed feet, but also in the innumerable small bone nodules that reinforced its skin into a veritable mail-coat. Such small bones occur in profusion in the tar, but the clue to their function comes from a very different part of the world: southern Patagonia.

Cueva Eberhardt, studied before the turn of the century by Erland Nordenskiöld of Sweden (the son of the famous discover of the North-East Passage), is a treasure trove of the past. In the cold and dry climate of this cave in southernmost Chile, not only the bones but also the dung and skin of great ground sloths, related to *Glossotherium*, have been preserved to our time. Such finds are very rare indeed. Preservation of hide and hair may result from freezing, as in Alaska. Certain types of bogs may also be conducive to such preservation. A third possibility is desiccation, as in this case.

The species found here, named for the nineteenth-century governor and explorer, Ramón Lista, is *Mylodon listai,* and belongs in the same family with the glossotheres. The patches of dried skin from the cave are indeed filled with pebblelike bones within the hide, which, when exposed, look "like a cobblestone pavement," says Professor William B. Scott. Among ground sloths, only the mylodont family sported this kind of mail. Incidentally, Lista, on one of his Patagonian expeditions, is said to have taken pot shots at a strange animal which scampered off without seeming to be hurt. It is tempting to think of Lista's bullet-proof quadruped as a nineteenth-century mylodont, but if so it must have been a young one, for it is described as much smaller than the fossil animals.

The skin shows that the mylodonts had a shaggy coat of fur, yellowish brown in color as preserved. The mylodon hairs have been put to some unexpected uses: Nordenskiöld found a very pretty bird's nest made out of sloth hair!

Man-made stone tools have also been found in Cueva Eberhardt, and there are radiocarbon dates of 10,000–11,000 years B.P. According to a somewhat adventurous theory, the early Indians kept the sloths shut up in the cave as a kind of living larder—but I don't think it should be taken very seriously.

Also in the cave there are great deposits of mylodont dung, which shows that the creatures fed on grasses, herbs, and shrubs; very probably, Harlan's sloth had a basically similar diet. An interesting discovery was the total absence of some important trace elements—copper, cobalt, and silver—reported by the Finnish geologist Martti Salmi. These trace elements were found in the dung of all the present-day animals which he studied as a control group. Salmi noted that at least copper and cobalt are necessary for the normal metabolism of mammals, and speculated that their lack in the food of the mylodonts about 10,000 years ago might have resulted in anemia, hastening the extinction of the species.

Whatever the cause, and whether Ramón Lista saw a living mylodont or something else, the North American glossotheres did not survive their Patagonian cousins. The youngest dated *Glossotherium* at Rancho La Brea lived some 13,800 years ago.

BIRDS

The record of birds from the tar is unusually rich. As we might expect, birds of prey are much in evidence. Vultures are common: they unfailingly notice animals in distress and would be among the first to react to the baiting of the trap. There are both extinct and surviving species among the vultures. While Old World vultures are related to eagles and other birds of prey, those of the New World are now considered to belong to the same group of birds as storks, herons, and ibises. This was also true for the extinct teratorns, of which one species, *Teratornis merriami*, was present at La Brea. It was the largest of the Brea birds, with a wing span of 11 or 12 feet. No living bird attains such a wing length (the albatross *Diomedea exulans* barely reaches 10 feet); yet much bigger teratorns are known, a Miocene South American species being the giant of its group, with wings spanning some 23 feet from tip to tip. But although related to the New World vultures, teratorns were clearly not scavengers but active hunters. Vultures have short, powerful beaks for ripping pieces out of large carcasses. The teratorn beak is very different, like that of a very long-billed eagle. It is now thought that teratorns preyed on small mammals which were swallowed whole.

Among the eagles, the golden eagle—which is a born scavenger—is predominant, but there are several other species, including the bald eagle, crested eagles of South American affinities, and the curious, long-legged, extinct *Wetmoregyps*. There are also falcons, hawks, and kites.

Extinct turkey *(Parapavo californicus)* from Rancho La Brea.

Among many beetles found at Rancho La Brea are burying beetles *(Necrophorus)*, here shown working on a dead chickadee.

As we have seen, an extinct turkey—called *Parapavo californicus*—is especially common in the tar, and we may assume that it was *the* game bird of the region. The presence of sheets of water is suggested by finds of ducks, geese, and wading birds. There are swans, storks, teals, plovers, avocets, snipes, curlews, cranes, and coots. But grebes are quite rare, and this probably means that there were few permanent ponds and lakes.

Pigeons are also found, including the passenger pigeon, which had vanished from the West in historical times, although it remained common in the north and east until its extermination in the nineteenth century. There are also owls, woodpeckers, and perching birds; even a partial list of the latter is overwhelming, reading almost like an incantation: larks, jays, magpies, ravens, crows, thrashers, waxwings, shrikes, meadowlarks, grosbeaks, siskins, goldfinches, towhees, sparrows, kingbirds, chickadees, bluebirds, blackbirds, orioles, etc.

Like the plant remains, these birds indicate a climate at times slightly drier than that of the present day, but not significantly colder. The cold of the ice-fields in the north did not reach southern California.

California quail *(Lophortyx californicus)*, a Rancho La Brea bird.

Californian horned lizard *(Phrynosoma coronatum)*, a Rancho La Brea reptile.

THE MAGIC WATER

Besides the tar pools, many other sites have preserved a record of Pleistocene life in the West; this is especially true of the caves. No story of cave exploration in California is complete without the tale of Samwel Cave in Shasta County, which was told by John C. Merriam many years ago.

His starting point was a Wintun Indian myth of a well with magic water, named Samwel. There were different versions of the story, but the end was always the same. Three maidens visited the well and drank the water, in the hope of improving their fortune. Disappointed by the result, they were told by an old woman of another well with greater power, much deeper in the cave. The search brought the maidens to the opening of a deep pit with sloping sides. One of them came too close, slipped on the moist rock, cried out, and fell. They could hear her scream; then she struck, and struck again, and all was still. The girls fled, and brought the tale back to their tribe.

Merriam and his party were attracted by the story of an unexplored cave. It might contain something of greater magic to them than water: fossil bones.

The location of Samwel Cave was known. It was in rugged country a few miles above the McCloud River. The explorers entered the cave and moved through galleries festooned with dripstone formations. The third chamber turned out to contain a well: perhaps the Samwel of the myth. On the floor lay fragments of bone deeply embedded in dripstone.

The second well and the deep shaft had not been found when the party returned to camp that evening. But one member, Eustace Furlong, stayed behind in the cave, and in the morning said that the shaft had been discovered and that ladders and ropes were needed. Dr. Merriam then brought a fifty-foot rope ladder and some loose ropes on a horse, but by then the shaft had been sounded at ninety feet. It was not until the next morning that all was ready for a descent. Burning paper was floated down the shaft, the presence of oxygen verified, and lots drawn. It fell on Furlong to climb down.

The chimney widened downward, but some yards down a protruding point of rock caused the climber considerable trouble. Still, at last Merriam could hear his friend's footsteps on the cave floor—and then a breathless exclamation: "There is a mountain lion down here!"

Now, sober judges consider this animal (more properly called puma or cougar) not dangerous to man—at least in daylight and in the open. A puma in a dark cave, possibly hungry and in a bad mood, is another thing altogether. No wonder Dr. Merriam made a split-second calculation of his chances to climb down and help his friend—and of the puma's chances to climb up and visit *him*. The next words, however, were reassuring—"It's a fossil mountain lion in the rock"—and then, in the same breath, "and here on the floor is the Indian girl."

Merriam descended, and they examined the remains of the poor girl. Only the bones and a thin organic film were preserved. In a few places, stalagmite crystals had started to form, but the time that had passed since she met her fate was too short for the dripstone to envelop her.

There were other things in the cave, things that had been there for a long, long time: skulls and bones of animals, many of them now extinct, and all covered with thick layers of stalagmite. The bones that Merriam and Furlong found here are now in the paleontology museum of the University of California at Berkeley.

One year later, the old entrance to the cave, through which all the animals had entered, was found; it was long blocked and forgotten. There were many nerve-racking moments in the exploration of Samwel Cave, but the one at which the pen seems to shudder in the serene scientist's hand refers to that first descent and to the fate of a unique bear skull. It belonged to the big Pleistocene black bear, and it was hoisted up the ninety-foot chimney in a bag that fell open a second before it could be saved. The skull was promptly transformed into a mass of debris on the cave floor. However, the skillful technicians at Berkeley did what all the King's horses and all the King's men could not do, and put it together again. It can be seen in the collection to this day, and a rare skull it is.

Merriam tells of his ecstatic walk around the museum on the cave floor, where the bones of extinct animals like the ground sloths, the mammoths, and the shrub oxen lay side by side with those of species still in existence, like the squirrel, the white-tailed deer, and the raccoon. And right there he made an astonishing discovery. Among the bones of the great extinct sloth he found a fragment of a human tooth!

Could Man have been here at the time of the sloths and mammoths? There was another possibility. With the tooth in his pocket Merriam turned back to the Indian maiden. A piece of a tooth was missing in her skull, and when Merriam tried to fit the fragment, the pieces suddenly seemed to melt together.

The riddle was solved; and at the same time, the ability of the Wintuns to hand down a tradition from generation to generation was verified. The scientist looked up through the chimney, where the protruding rock-point halfway up was outlined against the distant opening. "She struck, and struck again, and all was still." For a moment he relived it: she hit the point, a tooth fragment was struck off; then she plunged to the floor.

There was no trace of a well or of standing water on the floor. The girl had not found the water of greater power. But she was buried with great ceremony by the Wintun Indians at the point where the crystal waters of Nosone Creek join the Mc-Cloud.

SAN JOSECITO

From Shasta County in the north to Mexico in the south, caves, tar-pit sites, and river deposits give evidence of the great southwestern fauna of the Ice Age. As we move south, it gradually changes in character, and tropical forms start to show up. These lands were not troubled by the cold of the Ice Age.

A remarkable, and still incompletely studied, record comes from San Josecito Cave in Nuevo Leéon, Mexico. Already more than forty mammalian species, ranging in size from shrew to camel, have been identified in the immense ossuary, and more is to come. There are definitely Southern forms here—for instance, a vampire bat and a species of shrew now found only in southernmost Mexico. There are also old friends: Smiley the sabertooth is here, and so are the dire wolf, *Platygonus* the flathead peccary, and *Tremarctos* the Florida cave bear. Also, coyote, lion, jaguar, puma, bobcat, badger, and three different kinds of skunk have denned here—presumably at different times.

But then we also find some very foreign-looking beasts. One of them is related to the mountain "goat" of the northern Rockies, but much smaller: *Oreamnos harringtoni.* An even more surprising intruder in San Josecito is the typically Eurasian hunting dog or dhole, *Cuon.* We know it was present in Beringia, but no one dreamed it had gotten into America south of the ice until Ron Nowak of the University of Kansas came up with its telltale jaws and teeth, sorted out from a great mass of coyote bones. This is the red hunting dog of Rudyard Kipling's *Jungle Book.* Although his account

Mexican with antecedents in Asia: Dhole or red hunting dog *(Cuon)*

of the dhole's fierceness may be exaggerated, a score of hunting dogs may form a very efficient hunting pack. Their teeth, with well-developed shearing carnassial blades and reduced tuberculars, reflect the fact that plant food plays a much smaller part in their diet than, for instance, in that of the wolf. The presence of this animal once more points up the close relationship between the faunas of North America and Eurasia.

Another surprise came with the identification of the deer that is particularly common in the cave. There are thousands of bones and teeth of this animal, *Navahoceros fricki*, now extinct. Other remains show its presence in caves in other parts of Mexico, in New Mexico, and further north along the Rocky Mountains up to Wyoming. This was the mountain deer of North America.

It was truly a mountain deer. Its strong, stocky limbs were extremely adapted to a climbing mode of life like that of the present-day chamois and ibex. The only living deer which approaches it in this respect is the somewhat smaller South American huemul, or mountain deer of the Andes (rather incongruously named *Hippocamelus*).

Navahoceros males carried simple, three-tined antlers; probably a larger and bulkier head ornament would have been a serious encumbrance to a climbing form. The huemul antlers are still simpler, with just the two prongs, but lacking the brow tine of *Navahoceros*. George Miller now has evidence that *Hippocamelus*, too, once ranged into North America; but *Navahoceros* is by far the more common of the two. Quite recently Miller discovered an Irvingtonian forerunner in the Anza Borrego Desert.

We may imagine the beauty and daring of these supple animals moving about the cliffs and pinnacles of the great mountain range. Now they, too, have sunk into the earth whose pull they so bravely defied.

IN EXILE

The sight of an island on the horizon has always been a lure to adventurous spirits. And not only to man.

Four species of land mammals are found today on the Northern Channel Islands, separated from the Californian coast by the Santa Barbara Channel. They are small mammals: a harvest mouse, a deer mouse, a spotted skunk, a gray fox. There are also two species of salamanders, three lizards, and three snakes. Most of these (also two now-extinct species of deer mice) probably got to the islands by means of so-called waif dispersal: accidental rafting on floating trees and the like. Some, in fact, may have been brought out, wittingly or unwittingly, in Indian canoes. But one has to draw the limit somewhere. Phil C. Orr, reasonably, drew it at elephants: "While we can easily imagine an Indian lad taking a pet fox (as they did with dogs) in a canoe to the islands, it stretches the credibility to imagine the importation of elephants by that means."

Yet there were elephants—or, rather, mammoths—in the islands. Found at first on Santa Rosa Island in the 1870s, they have since been discovered on San Miguel Island too. They are known as exiled mammoths, *Mammuthus exilis*, and their presence here remained for long a much-discussed problem.

The exiled mammoths are very closely related to the great Jefferson mammoths of the continent, and in fact are now regarded as no more than an island race, a subspecies, of *Mammuthus jeffersonii*. Like their mainland cousins, they have very advanced cheek teeth with closely appressed plates, and so it seems that they were descendants of late Pleistocene mainland forms. In other words, they probably reached the islands during the last glaciation. They are smaller though. The shoulder height of the mainland mammoth was eleven feet or more, while the exiled mammoth reached about eight feet at most.

Two questions arise. How did they get to the islands? And why are they so small?

In answering the first question, we are reminded of the effect of continental glaciation on sea level. It dropped by three or four hundred feet. However, this does not create a land connection. To be sure, the lowering of the sea must have created a superisland, called Santaerosae, extending from San Miguel in the west to Anacapa in the east. Yet the deeper part of the Santa Barbara Channel remained submerged. And indeed we may reflect that, *had* there been a land connection, we would not have had just this haphazard collection of land animals on the islands today. Many more species would have made their way to the islands.

So the only possible answer (as noted, for instance, by Donald Lee Johnson) is that the mammoths got there by swimming.

The fact is that elephants are excellent swimmers. Although almost completely submerged in the water, so that only the top of the head and a small part of the back are visible, they use their trunk as a snorkle. At Trincomalee Harbor in Sri Lanka, elephants have been seen (and even filmed) swimming across a channel half a mile wide, of their own volition. In India, elephants have crossed the Ganges, taking up to six hours without touching bottom. And perhaps even more startling: in 1856, an elephant lost from a ship thirty miles out in the Atlantic off South Carolina was reported to have made its way to the shore in a heavy gale.

The distance from Fort Hueneme to Anacapa Island is about eight miles. Twenty thousand years ago, at the height of the last glaciation, the coast-to-coast gap was less than four miles. An elephant could swim this stretch in four hours or less. But why should it be tempted to do so?

Donald Johnson has an answer to that. Imagine a mammoth bathing in the surf zone off present-day Oxnard or Ventura—then about 12–13 miles further seaward than now. Behind it is a rather flat coastal plain, perhaps afflicted by drought, which led to food shortage. Before it looms the mountainous land mass of the great Santaerosae superisland, and the sea wind is bringing tempting odors of its vegetation. Elephants have poor close-up vision, relying on their trunk and sense of smell to explore their immediate surroundings; but there is some evidence that their long sight is good. So the Santaerosae would be a conspicuous landmark to the bathing mammoth, as well as promising good forage. And a few hours' swim would get it there.

But why are the exiled mammoths smaller than the mainland form which gave rise to them? Has the same thing happened elsewhere?

Yes, it has. We have seen already that the Caribbean ground sloths became dwarfed. But this has happened to elephants as well. Several Mediterranean islands became populated by elephants in the Ice Age, and some of these became incredibly small. In the cave deposits of Malta and Sicily, the process of dwarfing may be traced, stage after stage, ending up with a pony-sized creature only about three feet high. In the East Indies, too, insular proboscidians were dwarfed. Indeed, it is a general rule that large land mammals marooned on islands tend to become smaller. There were pygmy deer and pygmy hippos on the Mediterranean islands, and even today, a dwarfed form of the whitetailed deer is found in the Florida Keys.

There are various reasons why small size would be a selective advantage. In the first place, the insular terrain is often craggy, and agility is at a premium. Smaller size makes it easier to move around. But there are other and perhaps subtler reasons. By getting smaller you reduce your claims on food and living space, and both are limited on an island. As a result, the number of individuals may increase: instead of a small number of giants, the island can support a sizeable population of medium-sized animals. That is useful, because the dangers of inbreeding are reduced: there is a greater fund of genetic variation to draw upon. Also, there is less risk of a population becoming extinct by sheer accident.

These are, so to speak, positive or "active" factors of selection. There is a "passive" factor too: the absence of dangerous predators. The protection afforded by large size is no longer necessary.

The smaller size, then, may be seen as an adaptation to the insular environment. It may be thought that, later on, other mammoths swam the Channel. But once the island race had established itself, the intruders would be at a disadvantage. In effect, "this is our home: you do not belong."

Of course the reverse could happen—an exile returning to the mainland. The principle is the same: the exile would be inferior to the big mainland mammoths. But there is a suspicion that the exiles were not as good swimmers as their ancestors. With the dwarfing, the buoyant pneumatization of their skulls was probably reduced. And so, we may imagine, a status quo was reached; and they lived happily "until then took them the Destroyer of Delights and Desolator of Dwelling-Places" . . . some ten thousand years ago.

VIII

Mastodon Land

THE HUNTING OF THE MASTODON

TLASCALA, LAND OF GIANTS!
Amongst the marvels seen by Cortéz and his men in Mexico, and duly recorded by his chronicler, Bernal Diaz, there was the bone of a giant. It was a gift from the Spaniards' allies, the Tlascalans, rivals of the Aztecs. The bone was sent back to Spain to be viewed by the King, as evidence that this country was once the home of giants. That is what men believed in 1519, and what they continued to believe for many years.

The bone has since been lost (probably King Charles I of Spain, alias Charles V, Roman Emperor, cared less for it than for the gold of the Aztecs), but it was almost certainly the thighbone of a mastodon. Bones of these creatures still weather out of the rocks in the river valleys of Tlascala. But it is only recently that this has been recognized and the Mexican giant has been unmasked.

Indians thought the fossils were bones of giants who lived in the bowels of the earth and died if they came out into the sun—otherwise how could it be explained that they were never seen alive? Similar myths arose around the Siberian mammoths. It was not until the early eighteenth century that the first correct identification of a fossil American mammal was made, and this was not done by the learned men of the time—they tended to rave about monsters and giants—but by African slaves in California. They saw fossil mammoth molars and correctly recognized them as the teeth of elephants. Perhaps for a moment they had a vision of their old country.

In the year 1739, Charles Le Moyne, second Baron of Longueuil, set out from Montreal against the Chickasaws. His journey took him down the Ohio, and on the southern bank of that river, not far above the rapids at present-day Louisville, his expedition discovered mastodon teeth and bones. The exact site is not known, but it could be the place later known as Big Bone Lick, a salt marsh in Boone County, Kentucky, where great numbers of fossil bones of the mastodon and other Ice Age animals have since been dug out.

The Ohio Mastodon played an important part in the early history of vertebrate paleontology. Longueuil took his collection to Paris, where it was studied by the naturalist Louis Jean Marie Daubenton—who identified the tusk and thighbone as elephant but was unable to determine the molars, which are quite different from those of true elephants. Later on, they were seen by Georges Cuvier himself, the founder of vertebrate paleontology. He coined the name *Mastodon americanus;* the name mastodon (nipple-tooth) alludes to the shape of the molars with their paired, breastlike cusps. Cuvier also knew other kinds of mastodons, which had lived in other parts of the world, but *Le grand mastodonte de l'Ohio* was the biggest of them all.

Unfortunately, earlier naturalists had confused the mastodons with the mammoths, and so it happens that there is an earlier scientific name for the same animal: *Mammut americanum.* According to the priority rules, that is the name that has to be used in scientific language, though of course the mastodon is a quite different animal from the mammoth. Mammoths were real elephants, rather long-legged and with very massive cheek teeth, millstonelike structures which consist of many closely appressed enamel plates, adapted to chew the tough grass of the open plain. The mastodon, like the elephant, had a long trunk and big tusks, but it had shorter legs and its cheek teeth were simply constructed, with only a few cusps arranged in pairs and joined by cross crests. Such teeth are good for browsing but cannot cope with a grass diet. They are a heritage from the earliest ancestors of mastodons and elephants, and the mastodons remained conservative because they kept to softer food. Loris S. Russell calls the mastodon "a poor relation . . . never quite discarding the traces of his humble origin."

The mastodons had large, elephantlike tusks in their upper jaws. Some individuals had small tusks in the lower jaw, too (the lower tusks were quite large in some earlier species of mastodons), and Russell thinks they were used in feeding off the trees. The mastodon would pull a branch through its mouth, pressing it against the lower tusks with its tongue, and stripping off the leaves, twigs, and catkins. The stems of the branches would gradually wear a groove across the tusks just at the gumline, and this can be seen on the fossils.

Mastodon fossils are found all over North America, but they are particularly common in the eastern half of the continent, from Florida to eastern Canada; so this seems to have been their true homeland. John E. Guilday thinks mastodons favored stream and lake margins and swamps, the type of environment in which moose is found today. His suggestion recently received unexpected support when Kurt F. Hallin of the Milwaukee Museum described a find of mastodon skin and hair. (Earlier finds assumed to be mastodon hair have turned out to be algal strands.) The pelage consisted of fine guard hairs and a furry undercoat, very much like that in various aquatic and semiaquatic mammals. So, with Guilday, we may well think the mastodon was as fond of water as the moose. What a wondrous sight!

The mastodon was a very popular subject in the late eighteenth and early nineteenth century. George Washington acquired a mastodon tooth; and Thomas Jefferson, who did not accept the idea that any kind of living being could have become extinct, was sure that the species must be in existence still in the Northwest. Taking their cue, lesser spirits developed some amazing ideas of the appearance and habits of the mastodon. One author, carried away by the appearance of the tusks, even regarded it as a carnivore, and speaks of "this monster, with the agility and ferocity of the tiger. . . . Cruel as the bloody panther, swift as the descending eagle, terrible as the angel of right." Sane and judicious as always, Benjamin Franklin pointed out that the teeth of the mastodon clearly showed it to have been a vegetarian.

Soon, traveling impresarios were exhibiting mastodon skeletons in America and abroad. The London *Times* in 1841 published a notice of a skeleton put on show by Mr. Albert Koch, who had found it three years earlier in Gasconade County, Missouri; he called it "the great Missourium." (Other popular names for mastodons include The Great American Incognitum, The Leviathan Missourium, and the Carnivorous Elephant.) The notice reveals some skepticism as regards "our American friends," who "deal too much in the marvelous." The skeleton is described as "the fossil remains of a gigantic animal between whose legs, it is said, the mammoth may have strutted with ease. . . . The animal is supposed to be aquatic in its nature. This we should have inferred from the anatomical structure of its neck." Evidently, Koch's restoration was somewhat eccentric, and the last sentence may remind us that Mr. Koch became famous for exhibiting an enormous whale skeleton, too. The whale in question was an early Tertiary species (thus vastly older than the mastodon) which, incidentally, also has a curiously misleading scientific name. It is *Basilosaurus* or "imperial lizard"—its first describer happened to be completely mistaken about the affinities of the creature (whales are neither fish nor lizards but mammals). These whales reached a length of some twenty yards, an impressive enough figure, but that didn't seem good enough to Koch, so he combined the vertebrae of several different skeletons to produce an abomination three hundred feet long. He called it *Hydrarchus*—the ruler of the sea.

With this in mind, we can guess that Koch's mastodon left a great deal to be desired in the way of correctness. In fact, one Philadelphian, James Pedder (the editor of *The Farmer's Cabinet and American Herd Book*), thought the restoration so erroneous that he wrote the leading British anatomists and suggested various corrections. But he strayed even wider of the mark than Koch: "I have been led to conclude that the animal was a Monster of the Tortoise Tribe 32 feet long and corresponding width, with the power of withdrawing its head within its shell; the tusks then forming

Pedder's reconstruction of the "Missourium," after a copy by the British anatomist William Clift.

a mail of defense around its edge to ward off obstruction." He also made a sketch to show his tortodon, or mastoise, with the tusks as a kind of Salvador Dali moustache—they were "carried near the Earth, and resting upon it, at the will of the animal." It would probably be difficult to invent anything less like a mastodon than this product of Mr. Pedder's fertile brain.

At the same time, we should in fairness remember that even learned men knew very little about elephants at this time, and so could not be expected to recognize their bones. As G. G. Simpson notes, few Americans or Europeans had then seen elephants in the flesh. Koch's Missourium is now on display in the British Museum (Natural History), of course in properly reassembled condition. But mastodon skeletons are also displayed in many American and Canadian museums.

In spite of Jefferson's opinion, the mastodon is indeed extinct today. How long ago did the last mastodon die? There are some radiocarbon dates which have suggested the tribe may have survived long after the end of the Ice Age, perhaps as late as 5,000 years ago. But the dates are doubtful. The best authorities consider that such dates result from contamination with younger carbon; even a very small amount of intrusive live radiocarbon will result in a spuriously recent date. The accepted dates for late mastodons range from 9,000 to 12,000 years B.P.

There is now good evidence that early man in North America hunted mastodons as well as mammoths toward the end of the Wisconsinan, and this will be recounted in chapter 9. He may actually have made pictures of mastodons. A puzzling discovery was made in 1864, when H. T. Cresson and W. L. de Suralt found some artifacts near the Holly Oak railroad station in northern Delaware. Among them was a pendant made out of a whelk shell, with holes bored for a thong, and bearing the engraved image of a hairy proboscidean. It has been interpreted as a picture of the woolly mammoth, to which the peaked head shows some resemblance, but the shape of the tusks, the long, humpless body and the short legs are more reminiscent of the mastodon. Also, it is uncertain whether the woolly mammoth ranged as far south as Delaware. Unfortunately, the circumstances of the find are not clear; it was reputedly discovered with peat which had been dug for fertilizer, and was associated with a mixed bag of Paleo-Indian and younger artifacts. Most investigators tend to reject it as a hoax. But recent reexamination, at the Smithsonian Institution, indicates that the carving incisions show the same stage of weathering as the shell surface itself, suggesting that the picture is genuinely old. But how old? We are left with a tantalizing possibility.

In any case it is clear that the North American mastodon died out long ago, so long ago that any myths or traditions that may have formed around this strange, great creature have been utterly lost. But could mastodons have survived elsewhere? The discovery of living flat-headed peccaries should give us pause. In the 1920s, the German archaeologist K. Th. Preuss discovered a remarkable stele in the highlands of Colombia. About ten feet tall, it shows a man carrying on his head a human figure with an animal's head. That head has curved tusks, sticking out each side of a short trunk. Could it be the head of a mastodon?

Professor Ernst Stromer, the Munich paleontologist, did not exclude the possibility. Other suggestions range from pig (but there were no pigs in America in pre-Columbian days) to coatimundi; but the nose of the coati is not widened at the end like that of the figure, and of course this little carnivore does not have such tusks. On the other hand, the Indian artists often exaggerated the eye-teeth of their subjects, for instance in the man forming the lower part of the statue.

Even if this picture should in fact show a mastodon, it probably was a different species, for the American mastodon did not range into South America. Other species and genera took over there and they, too, are now gone.

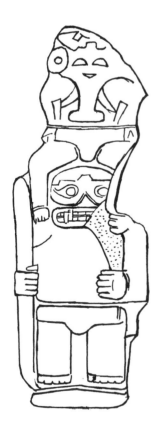

Pre-Columbian stele from Colombia, showing a man carrying a human figure with the head of animal. The head seems to have tusks and a proboscis. (After Stromer)

MOOSE: PAST AND PRESENT

As noted above, mastodons were especially common in the east. Their remains have even been dredged from the continental shelf, in the area that was exposed by the glacial regression.

Muskeg, a term of Cree Indian origin, is often used to denote a type of country now common in Canada. It is rather flat, traversed by very slow-moving streams and covered by peat with hummocks and stagnant pools; the forest consists of tamaracks and fir-trees. The term may also be applied to peat-bogs, swamps, and marshes in general. This kind of environment was apparently common in the eastern states during late Rancholabrean times.

Muskeg is typical moose habitat, and we might expect to find many remains of this great cervid accompanying the American mastodon during the Ice Age. But we do not, and the reason seems to be that moose did not exist in America south of the ice until at a very late stage in the Wisconsinan, when melting of the ice sheets had opened a way of immigration from Beringia.

The place of the moose was taken by another animal which has been called the stag-moose: *Cervalces scotti*. In size and in the build of the body and limbs, this animal compares very closely with the moose. But the head and antlers are quite different.

The moose head has a characteristic profile with its mobile, broad, and overhanging muzzle, almost like a rudimentary proboscis. The mobility of the muzzle is reflected in the skull with its greatly shortened nasal bones, which leave a large nasal opening. In *Cervalces*, on the other hand, the nasal bones were long and the muzzle presumably looked very like that in "ordinary" deer, such as the mule deer, for instance. It is of interest to note that the earliest Old World mooselike forms, from which the stag-moose had probably descended, had similar long nasal bones. The *Cervalces* line must have split off from the moose line (the genus *Alces*) before the evolution of the mooselike muzzle in the latter.

If the stag-moose had an "ordinary-looking" head, its antlers were extraordinary. They are hard to describe. Each antler has small palmations, numerous tines, and what Professor W. B. Scott (for whom the species was named) describes as "a great trumpetlike plate of bone on the lower side." It is, rather, as if a moose antler had been frantically torn and twisted, each tear sprouting new great tines. Deer antlers are organs of display: they serve as recognition signals and to impress rival males. One feels that those of the stag-moose must have been successful on both counts.

The ancestors of *Cervalces* doubtless immigrated from the Old World. When this happened we do not know. There are some finds of late Irvingtonian or early Rancholabrean age which may be stag-moose, although they are too fragmentary to give much

Stag-moose *Cervalces*, with the body of a moose but a normal deer's head, without an overhanging muzzle, sported great and intricately patterned antlers. Its extinction seems to coincide with the immigration of moose.

information. The oldest find comes from Nebraska. In the Sangamonian interglacial, the species ranged into Ontario. Wisconsinan records are numerous: the stag-moose ranged from Arkansas in the south to Michigan in the north, and from Oklahoma, Kansas, and Nebraska in the west to Virginia, Pennsylvania, and New Jersey in the east. It was still in existence 10,230 years ago; the date comes from a peat bog near Ansonia in Darke County, Ohio. Among the other material, excellent skeletons from peat-bogs in New Jersey are notable.

The extinction of *Cervalces* may have been hastened by competition with the true moose, *Alces alces.* Remains of moose have been found in the latest Wisconsinan and Postglacial deposits in Canada and in the eastern U.S. down to South Carolina. The range is much like that of the stag-moose, but the two are not found together, and the suggestion is that the moose took over where the stag-moose left off.

So the moose is really one of the most recent additions to American wildlife. Its origin was in Eurasia, where it is still present in the northern forests: it is a circumpolar species. A series of ancestral forms are known in the Old World. In mid-Pleistocene times there lived a species called the broad-fronted moose, *Cervalces latifrons;* some incomplete specimens from Beringia probably also represent this form. Its antlers differed from those of modern moose by having longer beams and smaller palmate parts. By Wisconsinan times Beringia was inhabited by the modern type of moose and the opening of the "corridor" made it possible for it to invade the American continent.

Another great member of the deer family, the wapiti or American elk, has a longer history in America. Present in Beringia since the Irvingtonian, it extended its range into the United States in early Rancholabrean times. There are a hundred-odd fossil occurrences in the north, northwest, and east down to North Carolina, Arkansas, and

Oklahoma. Within most of this vast range, the wapiti survived to historical times. Again, this is a circumpolar form; the American wapiti is very closely related to the Eurasian red deer, *Cervus elaphus.*

A third living species, the white-tailed deer *(Odocoileus virginianus),* is even more common and more widely spread in the fossil state than the wapiti. There are more than 150 certain occurrences and a lot more which have only been determined as to genus and so might represent mule deer *(O. hemionus).* As we have seen, the white-tailed deer is a very old species which has remained virtually unchanged since the Blancan.

In addition to these there was another species, now extinct, in the central-eastern parts of the United States east of the Great Plains. This is the sangamona deer, *Sangamona fugitiva.* A skeleton (and part of one) was found in a cave near Frankstown, Pennsylvania; unfortunately it is that of a female, and so the antlers are unknown. In addition, there are fragmentary finds from Illinois, Iowa, Maryland, Missouri, and Tennessee. The sangamona deer was somewhat smaller than the wapiti, but larger than a mule deer, and had exceptionally long, stiltlike legs. It must have been remarkably fleet of foot, and probably inhabited more open country than most cervids. A dated occurrence in the Brynjulfson Caves of Boone County, Missouri, suggests it survived at 9,440 B.P.. Now, like the Rocky Mountain deer *Navahoceros,* it is gone.

THE APPALACHIAN CAVES

The Appalachian mountain chain extends from New Brunswick to Alabama and Georgia. It is vastly older than the young mountains of the west. It started to build up near the end of the Paleozoic Era, a quarter billion years ago, in connection with the closing of a proto-Atlantic sea which brought America into touch with Europe. The chain continues in the mountains of Scotland and Scandinavia on the European side, now ripped apart from their American sisters by the subsequent opening of the modern Atlantic. Since then, the Appalachians have passed through many phases of erosion and rejuvenation. At one time, almost the entire range had been worn down to a flat plain, with only some of the highest peaks remaining; Mount Monadnock in New Hampshire gives an idea of what these solitary mountains may have looked like. Since then, upwarping and erosional dissection have recreated the ancient mountain chain, but some remnants of the old level surface can still be seen as the mountains flatten off at about 4,000 feet.

Appalachia creates an almost unbroken avenue of highlands from the far north to the Great Smokies of Tennessee and North Carolina. It brings a wedge of northern environments, fauna and flora in between the Great Plains in the west and the Florida-Carolina lowlands in the south and east. It is a world of its own; and so it was in the Ice Age. The higher ranges were glaciated and in the valleys paraded caribou, muskoxen, and woolly mammoths.

Much of the record of ice-age Appalachia has been extracted from the numerous fossil-bearing caves by a devoted band of explorers under the inspiring leadership of the late John E. Guilday of the Carnegie Museum of Natural History in Pittsburgh. Some of the caves date back to Irvingtonian times. But there is a group of caves which preserve a record of the late Wisconsinan and the transition to recent conditions.

The story can be read, for instance, in the 30-foot column of deposits in the New Paris No. 4 sinkhole, situated on Chestnut Ridge east of the Appalachian Plateau in Bedford County, Pennsylvania. It contained 2,700 vertebrate remains which, together with the pollen spectra, chart the evolution of the environment from a taiga parkland forest of 11,300 B.P. to boreal and temperate forest in post-Wisconsinan times.

The early levels contain high arctic forms such as the collared lemming, *Dicrostonyx hudsonius,* and the yellow-cheeked vole, *Microtus xanthognathus.* Of these two species, the lemming now occurs only in the tundra of Ungava Peninsula, east of Hudson Bay and north of the tree line; while the vole lives in the tundras, forests, and swamps of northwestern Canada and Alaska. There are only sparse finds of the lemming, but the yellow-cheeked vole is one of the commonest species in the Appalachian fauna of this period.

There are several other sites with this fauna. A most remarkable one is Natural Chimneys, a group of limeston пillars on the western side of the Shenandoah Valley in Augusta County, Virginia, at the base of which there is a shallow cave. The cave deposits yielded remains of 120 species of vertebrates—mammals, birds, reptiles, amphibians, and fish. An even more prolific site of the same age, also in Virginia, is Clark's Cave in Bath County, near Williamsville; no less than 143 species were represented here. Responsible for these tremendous accumulations of bones were owls and other birds of prey.

Many other caves of this type are known in other parts of Appalachia, in West Virginia, Kentucky, and Tennessee, and outliers of the same associations can be found in the numerous Ozark caves of Missouri. Pulaski County in that state, a good hunting ground for speleologists, is richer in known caves (212) than any other county in the United States.

Among the large mammals of Appalachia, many are already known to us from other areas. The big grazers, such as horse and bison, are scarce. Peccaries, on the other hand, are very common; most are *Platygonus,* but there are also some finds of a very different form, *Mylohyus nasutus* or the long-nosed peccary. This animal ranged over the eastern and central United States during the Irvingtonian and Rancholabrean, and an ancestral species occurs in the Blancan of Florida. It apparently was less common than *Platygonus,* except perhaps in Florida, and it might well have been described as a typical member of the peninsular fauna; but it did range north to Virginia and West Virginia, and occurs together with arctic lemmings and voles at New

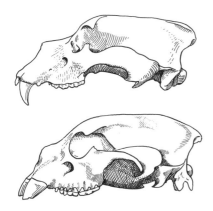

Skulls of grizzly (*Ursus arctos*, top) and great short-faced bear (*Arctodus simus*, bottom) show marked difference in length of snout.

Paris No. 4. The best-preserved finds, however, come from the scimitar-cat lair of Friesenhahn in Texas.

Mylohyus, close in size to a white-tailed deer, had a very long, slender snout and the slim and elongated legs of a fast-running animal. Its preferred habitat may have been the glades and forest edges, and it probably lived in about the same way as the Eurasian wild boar. An omnivorous animal, it shared this niche with *Platygonus* and the black bear—both very common in ice-age Appalachia. Of the three, only the bear survived in the end.

Some of the cave systems are extremely complicated, with long, branching passages, and it is no wonder that animals entering them sometimes got lost and finally died of thirst or starvation. There are several finds of lone jaguars, bears (among them the Florida cave bear), and other animals which have succumbed in this manner. Occasional footprints and scratches are poignant reminders of the agony of the trapped animals.

The great short-faced bear, *Arctodus simus*, and later on the grizzly bear, were also present in Appalachia. It appears that the grizzly spread into a great area toward the end of the Wisconsinan, far beyond that in which it has been recorded in historical times. And it may be that it crowded out its ecological predecessor, *Arctodus*, in its advance. There is only one place south of the ice in which the two species have been found together. This is Little Box Elder Cave in Converse County, Wyoming, a halfway station. The grizzly was present in Kentucky (Welsh Cave, Woodford County) about 12,950 B.P., and had reached southern Texas by 9,700 B.P. At Rancho La Brea, Los Angeles, it occurs in an early post-Wisconsinan fauna. Earlier faunas in Appalachia, Texas, and California have *Arctodus*, but not grizzly. The grizzly also spread into Ontario immediately after the recession of the ice.

Two kinds of ground sloths inhabited Appalachia. *Glossotherium* was described in the previous chapter. It remains to give a sketch of the Jefferson sloth, *Megalonyx*, the Great Claw originally described by Thomas Jefferson himself.

Mastodon Land   127

Megalonyx the Great Claw, Jefferson's ground sloth.

Could a science have a more auspicious beginning? Two papers on *Megalonyx*, read by the future president of the United States, and by Caspar Wistar, in 1797 before the American Philosophical Society, inaugurated the era of vertebrate paleontology in America. Jefferson at first had thought that the tremendous claw belonged to a carnivore of fabulous proportions. But a French paper on the South American *Megatherium* convinced him and Wistar that this was another member of the sloth tribe. Both he and Wistar used only the generic name, *Megalonyx,* without adding a trivial name. This was finally corrected by the French savant, A. G. Desmarest, who in 1822, most fittingly named the species *jeffersonii.*

Megalonyx jeffersonii was about the size of an ox, and only slightly smaller than *Glossotherium,* but much longer limbed. The great length of its arm would make it easy to pull down branches to browse on. Its hind feet were planted firmly on the ground, not turned inward as in most sloths, and the strong tail made the third leg of the familiar tripod. The head was short and broad. The fingers carried enormous claws, and the claws on the hind feet (the three central digits) were larger than in *Glossotherium* and the shasta sloth, and touched the ground in walking.

While *Megalonyx* has not been found in the Rocky Mountains or the Great Basin, it was present in the east and also along the west coast, where it spread into Alaska—the only ground sloth ranging into Beringia. It may have survived locally just beyond the end of the Wisconsinan; a terminal date from Evansville, Indiana, is 9,400 B.P.

Although known in South America since early Tertiary times, in the Ice Age megalonychid sloths were mainly distributed in North America and on the Antilles. The insular members of the family tended to be of small size, as in the case of insular mammoths, but the dwarfing in some cases was much more extreme, and some of the West Indian sloths were no larger than a dog. They reached the islands well back in the Tertiary, long before the first appearance of the family on the North American continent.

IX
Enter Man

THE ROAD TO AMERICA

IN THE PLEISTOCENE FAUNA of North America, animals of mixed origins intermingled. Many of them were old North Americans: tapirs, camels and horses, pronghorn antelopes, many families of rodents, and such carnivores as coyotes and short-faced bears. Others came from South America: water hogs, ground sloths, armadillos, glyptotheres. Some evolved in Eurasia and made their way across Beringia: muskoxen, bison, moose, and caribou. A few had their ultimate origin in Africa, for instance mammoths and lions. And in the last analysis, this also seems to be true for man.

As far as is known, all the human beings that ever reached America belonged to our own species, *Homo sapiens.* The species arose as a result of a long and complicated evolutionary history, which is outside the scope of this narrative. It is sufficient here to note that the earliest *Homo* evolved about two and a half million years ago in Africa, and spread into Eurasia about half a million years later. From these still very primitive members of our own genus, in a series of evolutionary advances which are as yet incompletely charted, men of the modern type arose perhaps about 150,000 years ago. It was long held that they originated substantially later and that *sapiens* evolved from Neandertal ancestors only about 35,000 years B.P. This appears now to be definitely negated by the discovery of undoubted sapients of greater age in Africa and Asia. In Africa, in fact, at least one modern-looking skull faces us across a time gap of 130,000 years. Much later, modern man reached Europe, where the oldest dated find presently known is about 32,000 years old. In Europe he probably coexisted with a rather more ancient race of Europeans, the Neandertals, for a thousand years or so, but by 30,000 B.P. the Neandertals were gone. The time-span of 40–30,000 years B.P. seems to have been one in which modern man spread vigorously into new areas: Sarawak in southeast Asia was reached at least 40,000 years ago, and Australia had human inhabitants 33,000 years ago, and possibly long before that.

Spreading overland, the only way for Ice Age man to reach the Americas was across Beringia. He could then establish himself on the American bridgehead, and advance to the south in the early stage of an interglacial or interstadial, when the land ice had melted sufficiently to leave an ice-free corridor between the western and eastern icefields. A second route which has been suggested would be along the coast, although this would mean bypassing some icefields reaching out into the sea.

In the orthodox view, the corridor route was used during the late Wisconsinan meltoff. About 13–12,000 years ago, the corridor was in existence, extending through western Canada from the Northwest Territories and Yukon boundary to the southeast, across Edmonton, Alberta, and opening on the Great Plains. Advancing through the corridor, man, like the grizzly bear and the moose (which apparently came at the same time), would have a great meltwater body on his left—Lake Agassiz—and the ice-clad Rockies on his right.

These men are called Paleoindians, and they were big-game hunters. Their characteristic stone tools are the so-called fluted projectile points, beautifully worked flint javelin-heads with an excavated base for fitting into a split shaft. The missiles may have been launched with the help of the throwing-stick or atlatl, still in use among the Eskimos when Fridtjof Nansen visited them around the turn of the century. And, almost as if by design, the Paleoindians have left some of these telltale points en route— in Alaska, at Old Crow River in the Yukon right at the gateway to the corridor, and near Edmonton.

From there on, advance was easy and Paleoindians rapidly colonized the continent from Florida in the east to Mexico and California in the west. About 11,000 B.P., a culture of mammoth-hunting peoples flourished in this vast area. Their distinctive tool was the Clovis point, a large variant of the fluted projectile point. Later on, perhaps when mammoths neared extinction, they were succeeded by primarily bison-hunting tribes who used the smaller, but also fluted, Folsom point.

This orthodox view of the colonization of America is strengthened by the fact that the Paleoindian cultures started to flourish slightly over 11,000 years ago, that is, at a time when the corridor had recently opened. Clovis itself, for instance—a site where several mammoths were killed while drinking in a pond in what is now New Mexico— is dated at 11,310 ± 240 years. There are, however, some dates which tend to suggest a somewhat earlier beginning, such as Wilson Butte Cave in Idaho, a lava blister where tools occur in a layer dated at 14,500 ± 250 years. Fort Rock Cave in Oregon is a runner-up, with a date of 13,200 ± 720 B.P. Does this mean that the corridor opened up earlier than thought? Or did these people arrive some other way, for instance along a coastal route? Or did the Paleoindian culture arise among people who had come to America at a much earlier date?

This last possibility will be discussed later on, but before that, there is more to say about immigration routes. Two have been mentioned, a "corridor route" and a coastal route. Both are based essentially on land travel. But did man in fact enter America on foot?

THE CASE FOR A SEA ROUTE

The third possibility, which has hardly at all been seriously considered, is arrival by sea.

When we think of early man moving from place to place we tend invariably to see him moving on foot. Yet, as Thor Heyerdahl has rightly pointed out, early man kept close to water. He tended to keep to the lake shore, the river bank, the seacoast. Millions of years ago, the australopithecines of Africa flocked to the lakes in the Great Rift Valley. Trekking overland, perhaps in rugged terrain and threatened by hungry carnivores and irate mammoths, is exhausting and dangerous. A dugout, a raft, even just a log to ride a-straddle, makes every sheet of water a comparatively safe highway. And so we may safely conclude that lakes, streams, and coastal waters played an important part in the migrations of early man.

All this is lifted out of the realm of speculation by a simple and incontrovertible fact: the colonization of Australia.

The earliest certain date for human occupation of this island continent is 33,000 B.P. It is based on mollusk shells, transported by man and found by Lake Mungo in New South Wales. But there are several other dates in the same time-range and older. Separately, they are less definite, but the total evidence, according to J. Peter White and James F. O'Connell, indicates that man reached Australia about 50,000 years ago.

Australia, together with New Guinea and a number of smaller islands, stands on the Sahul Platform, which was emergent in times of glacial regression. Similarly, the Indonesian islands of Sumatra, Java, and Borneo are on the Sunda Platform, which formed a great peninsula of southeast Asia when sea-level sank during the Ice Age. At the southeastern extremity of the peninsula, the present-day island of Bali was incorporated.

Between the Sunda and Sahul platforms extends the island world of Wallacea. It has been named for the renowned British zoogeographer and evolutionist, Alfred Russell Wallace, Darwin's contemporary and codiscoverer of the principle of Natural Selection, who did much of his epoch-making research in this area. The Wallacean islands stand in deep water, and so ensure the isolation of the Sahul lands with their aberrant, marsupial-dominated land fauna. Apart from the winged bats, the only land mammals of Asian origin to reach this area in prehistoric times were some forms of rats—and the wild dog or dingo, which was probably brought by man.

The rats probably got here by involuntary rafting. But man?

Even when the sea level was maximally depressed, there remain sea trips from landfast Bali eastward over the islands of Lombok-Sumbawa (then united), Flores, and Alor and then south to Timor. Gaps vary up to 14 miles (Alor to Timor). But from Timor southward across the Timor Sea to the Sahul edge, the distance is 54 miles at best, and probably even longer. The unknown land to be reached was low-lying and could not have loomed up on the horizon. So the early colonizers of Australia set out into what must have looked like the open sea.

Could man have made it rat fashion, helplessly clinging to a drifting tree? White and O'Connell very properly scout this idea. Human fecundity is low compared to that of rats, and for a colony to survive there must be a number of individuals of both sexes and appropriate age. Even such a promising group as three young women and three young men, with no incest taboo, stands a less than fifty percent chance of surviving. (They could, however, increase the chance by abolishing monogamy.) But somehow I do not see six young people of this description clinging in terror to a single tree, and drifting successfully across more than fifty miles of the open sea.

The only reasonable explanation seems to be that a colony was established quite deliberately; and this, in turn, indicates that Paleolithic men in the Sunda Archipelago must have been no mean seafarers. Whatever their crafts—dugouts, boats made from bundles of buoyant reed or the like—they must have practiced fishing off the Timor coast for generations, and in the course of such expeditions have sighted the Sahul coast.

Thor Heyerdahl, Björn Landström, and others have shown that human maritime traditions are far older than we used to think. The evidence for an early Timor Sea crossing extends it by more than a thousand generations and cannot be shaken by the subsequent degeneration of boatbuilding in Australia. Recent aboriginal watercraft in Australia are rafts and bark canoes not likely to endure the crossing. But traditions may be lost when the mode of life changes.

Sylvia J. Hallam, an Australian anthropologist and one of the few students who have seriously considered the possibility of a maritime entry into America, states: "Most hypotheses about the peopling of the Americas start by assuming, explicitly or implicitly, a dryshod crossing. . . . If Australia had already been reached by sea, this is unnecessary."

Remains of watercraft are necessarily perishable, and such remains of coastal fishing-based Ice Age settlements as may exist will normally be buried offshore beneath the high interglacial sea of today. So the direct evidence, if it exists, will be extremely hard to come by. All we can do at present is to scrutinize the indirect evidence at hand—for example, dates exceeding the acceptable limits for the "corridor route"—and, in general, keep an open mind.

Man reached Japan before the onset of the last glaciation, perhaps as early as during the penultimate one. A glacial regression would narrow the gap between Korea and Japan to sighting distance. The distance from Japan to the west coast of North America is awesome, yet the warm Kuroshio stream would speed a vessel that way in its great sweep across the North Pacific. Amassing an impressive amount of evidence, Thor Heyerdahl shows that men in a later age did cross just that way in primitive craft.

I shall leave it at that, with the addition that a combination of a sea and land route—that is, an advance along the coast, perhaps for many generations—may be a more likely alternative.

THE FIRST AMERICANS

Did the Paleoindians bring their traditions with them from Asia, or does their culture have deep roots in a still more ancient American past? Both alternatives have their advocates. The former is the orthodox one, and it reckons with a comparatively late entrance of man. There are, however, reasons for suggesting a much greater antiquity of man in America.

Numerous finds have been dated, by radiocarbon or other methods, much earlier than those enumerated so far. They have sparked many controversies, and in many cases the evidence has a tendency to dissolve. A number of human skulls and other bones from various sites in California were dated in the range of 17,000 to more than 40,000 years B.P.; but the dates have later been revised and turn out to be quite recent. Near Taber in Alberta, the fragile skull of a 2-year-old child was found in deposits which had later been overriden by the main Wisconsin ice sheet, over 20,000 years ago—but the skull nevertheless is Recent. And so is Otavalo Man from a cave in Ecuador, although originally dated at 28,000 B.P.

However, as Richard MacNeish has pointed out, there are some fifty sites in the Americas in which human bones or artifacts have been dated at over 12,000 B.P. There is, for instance, the find of a mammoth associated with tools and flakes at China Lake, some 60 miles north-northwest of Barstow, California, and dated at over 40,000 years. Used in dating were uranium and thorium from the enamel of the mammoth tooth. (The mammoth was an aged individual and so possibly was scavenged rather than hunted.) Remains of horses, camels, and small mammals were also recovered at the site, which was dug under the direction of Emma Lou Davies.

Another instance, from South America this time, is a rock shelter by the Piaui River in northeastern Brazil excavated by the French students N. Guidon and G. Delibrias, where there seems to have been continuous human occupation since 32,160 ± 1000 B.P. and up to 6,160 ± 130 B.P. These are radiocarbon dates from charcoal, and so should be very reliable, the more so as there is a sequence of 17 such dates!

Some of the paintings on the wall of the rock shelter are evidently of great antiquity as spalled fragments of painted rock are found in the very oldest levels.

The Old Crow River basin in the northern Yukon, studied by the Northern Yukon Research Programme of the University of Toronto, contains a sequence of superposed strata reaching back to the Illinoian. In these deposits, artefacts have been reported, as early as in the late Illinoian: they consist of bone tools with evident percussion marks. However, not all students will accept such evidence, and the same is true for the Sangamon-interglacial deposits near Medicine Hat in Alberta, in which innumerable flakes have been found. They must be about 100,000 years old. Are they man-made or produced by natural agencies? Opinions differ.

In the scenario proposed by MacNeish, the prehistory of the Paleoindians extends back to 70,000 B.P., give or take 30,000 years, and passes through a series of stages which are reflected in the development of the stone-tools. In the very early stages, man was still an unskilled hunter, and had but slight impact on the fauna. It was only in the final stage, that of the true Paleoindian cultures—13,000 to 8,500 B.P.—that man in America became a true big-game hunter.

Inevitably, all this has led to far-reaching speculations, even to the effect that our own species, *Homo sapiens,* really evolved in North America and thence spread to the Old World. At present this idea is certainly premature, and the evidence at hand does not support it.

Old ideas die hard. Most scientists have a conservative streak, and this is quite justified. Most newfangled ideas are in fact wrong. Yet some have substance, and their time will come. An open mind is a good thing to have—in science, as in other walks of life.

If man (in whatever guise) was in fact present in North America for such a long time, he left few traces and made curiously little impression on his world. With the advent of the Paleoindians, the picture changes completely. Whether this culture arose from an older one in America, or brought its traditions from Asian soil, it caused a tremendous change in the American scene.

DYING OUT

The Ice Age: great glaciers and large mammals. Now, both have dwindled. Continental ice sheets remain in Antarctica and Greenland, large mammals in some restricted areas. The waning of the ice sheets, though, has happened before—over and over again, in the course of the Pleistocene. The going of the megafauna, those great herbivores and the magnificent carnivores that depended on them, is a new thing.

As such, extinction is nothing new. New species evolve and old ones die out; evolution and extinction are the two sides of one coin. Still, the late Pleistocene story differs from earlier ones in at least three respects.

In the first place, there was no replacement. Species became extinct, but no other species immigrated or evolved to take their place. The fauna was simply impoverished. At the moment, I can suggest only two exceptions to this rule. One concerns the extinction of the great short-faced bear, *Arctodus simus.* It coincides with the immigration of the grizzly bear, *Ursus arctos*, which may well be the ecological successor of *Arctodus.* The other is the case of the stag-moose, *Cervalces scotti.* Its extinction might be due to competition with the invading true moose, *Alces alces.* But that is all.

Second, the vast majority of the species that died out were large mammals—those forms of animal life which, more than anything else, made the fauna of the Ice Age so spectacular. Mammoths, mastodons, ground sloths, glyptodonts, hosts of horse and camel species, pronghorns and peccaries, great carnivores like the sabertoothed cats, short-faced bears, and dire wolves: all gone. In contrast, the small mammals were hardly affected at all. Blancan and Irvingtonian extinction affected animals of all sizes.

Third, the time-rate of this extinction seems to be quite out of the ordinary. In the Blancan, Irvingtonian, and earlier Rancholabrean, extinction is spread out over hundreds of thousands of years. At the end of the Wisconsinan, on the other hand, its rate is whipped up to tenfold or even a hundredfold.

We can get some idea of its rate by looking at the youngest known record for each species. Naturally, this will not be a true terminal date, because the species may have lived on for an unknown period of time. However, if you cannot have the best (a precise expiration date), you must make do with second best. Unfortunately, even this sort of terminal date is only available for a limited number of species.

They do give some interesting information, however. The dates point to a great concentration of extinctions (or, more prudently stated, last known occurrences) in the time-span between 12,000 and 8,000 years ago. In the course of these four millennia, some 26 species of large herbivorous mammals (that is, megafauna) probably died out: six during the first, six during the second, nine during the third, and five during the fourth. The last occurrences range almost symmetrically around the date 10,000 B.P.—which is indeed the conventionally accepted date for the end of the Ice Age. During the same millennia, at least six species of carnivores, but no more than two species of rodents, died out.

These figures are minimum numbers. There are many species for which terminal dates are not available at present, and they may be expected to swell the roster. Also, future corrections will certainly affect the distribution of dates. It stands to reason that new information will tend to push some dates closer to our time. On the other hand, some of the dates may turn out to be incorrect as such, due for instance to impurities in the samples. This may be especially likely for the very young dates, so they may tend to be pushed backward in time. For my part, I suspect that rather more dates will be found to fall within the millennium between 10,000 and 11,000 B.P. Indeed, those at hand already tend to crowd within that time-span.

A millennium is a long time to us. Yet in the history of the earth and its life, it is only a fleeting moment, and in this light the wholesale extinction of so many species seems almost cataclysmic.

THE OVERKILL THEORY

What killed the megafauna?

Only two possibilities are taken seriously by most students. One is overkill by hunting man; the other is climatic change, with resulting environmental change. There are strong arguments for both. Let us start with a look at the overkill theory.

Modern man, of course, poses a serious threat to wildlife of many kinds. In the first place, there are so many of us that there is very little space left for others. Also, with the help of modern firearms, we have hunted many species to extinction or near-extinction—in North America, for example, the passenger pigeon and the bison, respectively—in a short time. Agriculture and urbanization carves slice after slice from nature. Pollution of water, air, and ground has its phantom echo in the hush of the silent spring. But all this belongs to our time. We may feel that the men of 12,000 years ago were too few, and too feebly armed, to devastate the wildlife of an entire continent.

On a smaller scale, even humans armed only with stone-age weapons have apparently succeeded in doing something of the kind. The island of Madagascar was colonized by man about 800 A.D., and in less than a thousand years, when the first European explorers came, its wonderful large mammals and ground birds had vanished. The story of the great moa birds of New Zealand is similar. In the West Indies, ground sloths survived long after their extinction on the mainland; here, too, the advent of man brought about their destruction. Most students agree that hunting man was the main factor in the extinction on such islands. (Of course, similar tragedies on isolated islands have been repeated time and again in later times, and the dodo of Mauritius stands as a symbol of all that is dead and gone: dead as a dodo.)

Paul S. Martin of Tucson, Arizona, chief advocate for the overkill theory, points to the almost invariable juxtaposition between advanced hunting cultures and mega-fauna extinction. An early episode of extinction in Africa, some 50-60,000 years ago, appears to coincide with the appearance of quite sophisticated hunting weapons. In Europe, the entrance of modern man some 32,000 years ago heralded megafauna extinction. In America the correspondence is especially clear: Paleoindians were at their peak around 11,000 B.P., and the great extinction occurred at the same time. There seems to be no possible climatic factor in common—only man.

The behavior of animals on isolated islands may also give a significant clue. These creatures avoid their natural predators, but are quite tame to man, who is a completely new phenomenon to them, and so are easily killed. The same was presumably

Ground sloth rearing up in defence position would probably intimidate approaching predator by its great height, but holds arms ready to take a swipe.

true for those American mammals which had no previous experience of man. This can be tested. If man was a significant factor, then the extinction must strike particularly hard at the endemic American species—those which had evolved locally for perhaps a million years or more. On the other hand, more recent immigrants from Eurasia would have descended from stocks with previous experience of man, and would have inherited appropriate flight reactions.

Such is indeed the case. The megafauna of the late Ice Age comprised some 49 species which had evolved there and over 90 percent of them (45 species) became extinct. On the other hand, of the 13 species which were immigrants from Eurasia, only 6 —less than 50 percent—shared that fate.

Interestingly, the difference between endemic and immigrant carnivores is much less—35 endemics, 11 of them extinct, or 33 percent; 10 immigrant, 2 extinct, or 20 percent. This may suggest that the prime factor in carnivore extinction was not the immediate influence of man, but simply the disappearance of their prey.

An important part of science is the developing of "models"—or scenarios, if you like—to explain the facts at hand. They are not intended to be anything like the "last word": they explore possible explanations, and they make suggestions for future research and predictions which can be put to the test. It is a good thing to have numerous alternative models. Many will turn out to be incorrect, in details or in their main structure, yet they are usually invigorating. One such scenario was worked out by Paul S. Martin, and it reads like a tremendous epic.

It begins with a small band of Paleoindians entering through the "corridor," south of Edmonton, Alberta. They find a virgin land with an enormous wealth of large, easily hunted mammals. Being already highly efficient big-game hunters, they make this game their livelihood. As they advance to the south, rapidly multiplying in numbers, they kill off the animals. No matter: there is always more game in front of them. They only need to advance about 10 miles a year, on average, and even though their numbers double every 20 years, there is always enough game before them to keep them going. In this model, 300,000 hunters wipe out 100 million large mammals in 300 years. And in that period they reach the Gulf Coast and northern Mexico, and for most of them, the good life comes to an end. But a small echelon presses on through the narrow Central American isthmus, and finds a new continent to sweep through.

Finally, a thousand years after the first entry, the late Paleoindians stand in southernmost South America—Ultima Esperanza, the Last Hope—and have nowhere to go. A self-destroying culture comes to an end.

Thus, the Paleoindians are pictured as sweeping across the continent, killing the megafauna on a broad front en route, and finally coming to a properly bad end, having destroyed the means of their own livelihood. It is an intensely dramatic model, the extreme of its kind, and one which gives modern man a lot to think about, for are we not doing much the same right here and now?

Of course, many were left behind, and they had perforce to modify their ways, and try to live in concert with nature, rather than by exploiting it. There are testimonials to the fact that Amerindians of historical times were deeply concerned with the balance between man and nature.

Paul Martin's model has met with severe criticism and is clearly at odds with some of the information on dating and cultural development. According to the model, at any given point the Paleoindians would sweep through very rapidly and then be gone, yet a sequence from Clovis to Folsom, spanning over thousands of years, can be traced. The model is probably too explosive to fit the facts. That, however, does not invalidate the concept of overkill, nor does it preclude the possibility that elements of the model may turn out to be highly useful.

An interesting comment on the overkill idea was recently made by Antti Järvi of Helsinki. He points out that a predator has to live in a dynamic balance with its prey, and that the numbers of both will tend to fluctuate in rapport with each other. However, Antti speculates, if you can augment your diet with plant food, the balance could break down. (In effect, vegetarianism rearing its ugly head!) Now, fossil excrements of early men like the Neandertals and their predecessors of Europe actually point to a purely animal-based economy with no trace of plant food. In contrast, late Ice Age modern man evidently ate plant as well as animal food. In that situation, you might kill off your big game without having to face starvation as a consequence. Here, again, is an example of a creative model which may stimulate further work.

KILL SITES

The known mammoth kills from the Clovis cultural period range in time from 12,650 B.P. (Lubbock Lake, Texas) to 11,160 B.P. (Domebo, Ohio), with a striking concentration in the last century or two. The Domebo date is also the terminal date for the Jefferson mammoth, so it probably became extinct at about this time.

The mammoths were killed at watering-places. Probably, this was the best situation in which such large animals could be hunted, though it should be remembered that remains from kills in other surroundings would stand less chance of surviving to the present day.

A good example is the Lehner Ranch in the upper San Pedro Valley, near Hereford, Arizona. There are other mammoth kill sites in this area, too. About ten miles to the north lies Murray Springs, where two mammoths were found, and a similar distance to the southeast one mammoth was killed at Naco. At the Lehner site, however, at least eight mammoths were bagged, probably at different times. Like other late finds of Jefferson mammoths, these animals were comparatively small—still, they must have been awesome beasts to hunt with Stone Age weapons.

The butchering took place on a sand and gravel bar at the southern bank of an ancient stream channel, which has been named Mammoth Kill Creek; it overlaps with, and is roughly parallel to, a present-day arroyo. In that place, beneath a high red clay bank, the creek formed a shallow, placid freshwater pool, where the animals would go to drink and wallow in the water. Thirteen projectile points and various other implements were found with the bones, and a little way upstream, the remains of two hearths (made at different times) were unearthed. In addition to the mammoths, fragmentary remains of bison, horse, and tapirs have been identified, but it is not known whether these were killed by man.

At Lubbock Lake in the northwestern part of the city of Lubbock, Texas, a long story of human occupation starts with a Clovis culture and a mammoth kill dated 12,650±350 B.P. At a higher level, dated 9,883±350 B.P., were found Folsom points together with remains of bison and the extinct pronghorn antelope *Capromeryx*. The record then continues with a sequence of more recent occupations. As a population center, then, Lubbock has a tradition which goes back almost 13,000 years, and puts many ancient Old World cities to shame!

With the mammoth gone, the bison became the most important game animal of the Paleoindians, and there are literally hundreds of bison kill sites in the Great Plains. Large herds with up to 200 or more animals were driven into traps and killed. Excellent natural traps of this kind were the head of an arroyo or the trough of a parabolic dune. An example of the latter is the famous site of Casper (Natrona Co., Wyoming), where the remains of about one hundred bison were found. In the trough of the dune there was a pond with lush vegetation. Among the bison, remains of one camel *(Ca-*

melops) were also found. This may have been an animal which mixed in with the bison herd, or it may have been hunted at another time; at any rate, its bones show butchering marks similar to those on the bison remains.

There is a second find of a butchered *Camelops* at another site in Wyoming; that one dates back to Clovis times. There are also many localities where *Camelops* remains and Paleoindian tools are both present, but evidence of butchering is lacking.

Whether the American mastodon was hunted by man has been debated for more than 150 years. There are now two or three sites where the evidence seems fairly conclusive. Perhaps the most interesting is Kimmswick, some 20 miles south of St. Louis, Missouri, close to the Mississippi. Two mastodons were killed here at different times, and among the bones of one of them, Clovis projectile points were found. Judging from associated small mammal remains, the area was one with deciduous woodlands and open grassy areas, a setting very different from that of the Clovis mammoth kills in the Southwest.

In South America, the role of the zygodont American mastodon was taken over by bunodont mastodons, for instance the genus *Haplomastodon.* These were closely re-

Life restorations of three Pleistocene bunodont mastodonts. Top, *Cuvieronius*; center, *Stegomastodon*; bottom, *Haplomastodon.* All lived in South America, *Stegomastodon* also in the Blancan and early Irvingtonian of North America. *Haplomastodon* is the only one known to have been hunted by man. (Based on restorations by Ibsen de Gusmão Câmara)

lated to the stegomastodonts which inhabited North America in the Blancan. Reconstructions of *Haplomastodon* show a medium-sized proboscidean with moderately large, upward-curving tusks. Some 13,000 years ago, human beings in what is now Venezuela slaughtered a haplomastodon calf. The site is Taima-Taima in the state of Falcón, east of the city of Coro. The site was then a much-frequented waterhole with a supply of artesian well water. A broken projectile point was found within the hipbone cavity, other tools were scattered around the carcass, and the bones showed butchering cuts where the tendons had been attached. The remains had been left as the animal had collapsed on its left side, but some parts had been dismembered and the hunters evidently carried off the head and the right foreleg. In addition to the calf, bones of adult mastodons, glyptodons, horses, sloths, and even a big cat and a bear were found in the same layer; however, what killed them is unknown.

It might be thought that the unwieldy ground sloths would have been excellent game for Paleoindians, but to date no definite evidence for sloth hunting has been found in North America.

Among Paleoindian kills, bison far outnumber the other animals. Yet the bison is still with us, and as far as we know, its existence was not endangered until the nineteenth century.

CLIMATE AND EXTINCTION

The extinction coincides with the time of the Paleoindians. It also coincides with the transition from the Pleistocene Ice Age to the Recent epoch. Could the extinction be a result of climatic change?

At first sight, the climatic history of the Ice Age seems to contradict the idea. The interglacial in which we now live is not the first of its kind. They have been recurrent, every 100,000 years or so, for more than a million years. There is even some evidence that the interglacial of 100,000 years ago was warmer than the one we live in now. Such, at any rate, was the case in Europe, where hippopotami and other southern forms ranged well into the British Isles. Yet it did not bring about any mass extinction of megafauna.

On the other hand, the climatic history of the present interglacial may have differed in some important respects from previous ones. There is, for instance, a suggestion that a period of unusual dryness occurred in an early phase. As John E. Guilday has shown, such a dry phase might work out as a bottleneck affecting the survival of large mammals. It might, for instance, result in the fragmentation and restriction of a given habitat. This would reduce the numbers of the species dependent on that habitat (both large and small). However, small animals generally need a much smaller individual range than large ones, and those that are left will be numerous enough to carry the species through the bottleneck. For the large species, the outlook would be less favorable, and might be further dimmed if species that were not previously competing for their livelihood were now thrown into competition. Thus the large species, demanding and utilizing great resources, would be more vulnerable than the small, unpretentious one.

In a somewhat different approach, R. Dale Guthrie, studying the "mammoth steppe" of the northern areas—in North America as well as in Eurasia—suggests changes of a nature which could account for much megafauna extinction.

In chapter 6, it was noted that some Pleistocene communities of small mammals were very different from those of the present day. Species which are not now found together, and sometimes live a thousand miles apart in our time, then coexisted in the same area, and their remains are found together in the deposits. The suggestion is that conditions were then different from anything found today. In Guthrie's opinion, the megafauna indicates much the same thing, and indeed such conditions—whatever they were—may have been necessary to support the megafauna.

The key word, according to Guthrie, is *variety*. The modern tundra, for instance, is very monotonous. Enormous tracts are covered by much the same sparse vegetation. On the other hand, the great plant-eaters of the mammoth steppe were dependent on a varied spectrum of food plants. Horses tend to specialize in medium-level grass stems with fairly low protein content. Bison also need a high fraction of grass sheaths and leaves. Sheep prefer new growth of high-quality herbs. Musk oxen feed on low deciduous shrubs. Mammoths, probably, fed on herbaceous vegetation in a rather unselective manner, but needed copious amounts of it—modern elephants are voracious eaters. All these animals with their varied appetites could not have coexisted on the modern tundra. They would require a more varied environment, which they could partition according to preference.

Guthrie suggests a mosaic-type environment with several different kinds of floral communities, rather than the sharply segregated zonation of today. One important factor, also, is the distribution of rainfall over the year. Precipitation in the north today is seasonal and this limits the growth season of the vegetation. With a more even distribution of the rainfall, even the low precipitation of the northern areas could bring about an increase in plant growth, and more specially, a lengthening of the time during which abundant forage is at hand.

This could explain the large size and high quality of the Pleistocene game animals. An Arctic herbivore grows only in summer, in that season when abundant and nutritious forage is at hand. Guthrie thinks that the Pleistocene mammals grew so large because that season of peak range quality was then longer. The more even distribution of moisture would result in a patchy environmental mosaic where different plants reached their optimum at different times. Preliminary tests on captive Dall sheep, made by Guthrie, support this idea.

Another characteristic of the mammoth steppe was that there was probably little snow cover. Many of the Pleistocene mammals had small hoofs—the horses, for instance—and so would be immobilized by deep snow (in contrast with, for instance, moose and caribou). Guthrie suggests that this was due to strong winds sweeping large tracts bare.

As a corollary, we may well suspect that a mammoth, for instance, would not be able to subsist on a present-day tundra. And, to be sure, the mammoths killed by Paleoindians are generally stunted in size, so perhaps a change in the productivity of the environment was already under way at that time.

In its own environment, the mammoth-steppe megafauna was a well-balanced community. The rigors of winter kept the numbers of animals in check, so that there was no overgrazing in spring, but plenty of food for all.

In quite another way, the importance of the diet factor is brought out by the fact that the surviving megafauna consists of ruminants—animals that have a complex stomach and chew the cud. Among those that died out, some ruminants are also found, but the majority—for instance horses, mammoths, sloths—had a simple stomach.

So a quite new and exciting theory of Pleistocene gigantism, and of the end-Pleistocene extinction, is now being developed—at least as far as the northern areas are concerned. It may turn out that analogous factors were at play in some southern areas as well. Perhaps a new and successful theory will combine such influences with that of man, appearing on the scene and delivering the final blow to already dwindling herds of big game. That could have been a rapid process: the Clovis mammoth kills seem all, or almost all, to have occurred within one or two centuries.

Whatever the explanation, the giants have gone to rest. Yet we are still left with a goodly Pleistocene fauna which has survived to our day. Its future fate is now our responsibility. If these pages and pictures have for a moment succeeded in bringing these phantoms from the past to a revenant existence, they may induce us to face that responsibility. The wildlife of North America should live long—not just in books, but in real life.

Epilogue: The Future

WHAT OF THE FUTURE? Will another glaciation be upon us, suddenly, within the next few decades or centuries? Assuredly not. The available evidence tells us that glaciations develop very slowly, with many oscillations between advance and retreat; if anything, they end at a faster rate than they begin. Still, the Astronomical Theory and the record of previous glaciations suggest that we are gradually moving into a shadow zone foreboding the end of the present interglacial and the beginning of a glaciation which should attain its maximum some 23,000 years hence.

The warmest part of the present interglacial, called its climatic optimum, was reached about 7,000 years ago. Since then the major trend in temperature has been downward, but superimposed upon that trend are shorter oscillations with a length of about 2,500 years. The last of these caused the so-called "Little Ice Age," lasting from 1450 to 1850 A.D. The causes of these changes are not well understood at present. Still, it might be expected that future "little ice ages" of increasing severity will occur, leading finally to truly glacial conditions.

All this is based on the assumption that Man does not meddle. But Man *is* meddling by burning great amounts of fossil fuels, and so increasing the amount of carbon dioxide in the atmosphere. This brings about a hothouse effect, leading to a rise in the global temperature. The effect has been variously interpreted. It could well bring about a "super-interglacial," distinctly warmer than the Sangamonian, before the onset of future glaciation. Such a warm period might result in partial melting of present-day continental ice sheets and a corresponding rise of sea levels, inundating low-lying areas. Still, unless the climate system of the earth is fundamentally affected, the cooling trend would only be postponed, perhaps to reassert itself about 2,000 years from now.

The Ice Age will probably return. The great faunas will not. They are lost forever, and it would take hundreds of thousands, perhaps millions of years for evolution to produce a new megafauna. Optimists have toyed with the idea of reviving frozen cells from mammoth cadavers and using its genetic material to have an elephant cow bear a mammoth baby. Unfortunately, even in the best-preserved frozen mammoths, most of the proteins have deteriorated, and no reviving is possible. The mammoths and ground sloths, the saber-tooths and glyptotheres will remain shadows in the past. So let us treasure what we have left.

Suggestions
for Further Reading

THERE ARE SEVERAL scientific journals devoted entirely to the study of the Quaternary period (which includes the Pleistocene epoch or Ice Age, and the Holocene or Recent epoch), for instance the *Journal of Quaternary Research* (Academic Press), *Quaternaria* (Rome, Italy) and *Boreas* (Oslo, Norway). Yale University publishes the journal *Radiocarbon*, which lists C-14 dates on a worldwide basis.

Reports on Ice Age life are scattered in many kinds of journals. Access to this literature is now facilitated by the book *Pleistocene Mammals of North America*, by B. Kurtén and E. Anderson (Columbia University Press, 1980) in which more than 250 Blancan, Irvingtonian, and Rancholabrean sites, and all the 562 mammalian species known in the fossil state, are described: it has been the main source for the present book. Its bibliography, with over 1,000 entries, will lead the serious student to the pertinent literature on most of the topics treated here. A worldwide catalogue of fossil birds by P. Brodkorb has been published in instalments in the *Bulletin of the Florida State Museum*, starting in 1963. W. Neill's *The Last Ruling Reptiles* (Columbia University Press, 1971) tells of today's crocodilians and their predecessors.

The natural history of North America, including its geological past, is treated in the National Geograph Society's magnificently illustrated *Our Continent: A Natural History of North America* (1976, ed. S. L. Fishbein). A classical textbook on the Ice Age is R. F. Flint's *Glacial and Quaternary Geology* (Wiley, 1976). An absorbing account of the history of Ice Age research and the causes of the Ice Age is J. and K. P.

Imbrie's *Ice Ages: Solving the Mystery* (Enslow, 1979). *The Ice Age,* by B. Kurtén (Putnam; 1972) is a popular account, while the great volume *The Quaternary of the United States* (ed. H. E. Wright and D. G. Frey), put out by the Princeton University Press for the Congress of the International Association for Quaternary Research in 1965, is highly technical.

Many books deal with fossil mammals and evolution. The classic is G. G. Simpson's *Tempo and Mode in Evolution* (Columbia University Press, 1944), which was to revolutionize evolutionary paleontology and, in retrospect, stands out as the single most important book of the century in this field. The standard textbook is A. S. Romer, *Vertebrate Paleontology* (University of Chicago Press, 1966). For less technical yet profound views on evolution, see, for instance, L. Eiseley, *The Immense Journey* (Vintage Books) and S. J. Gould, *Ever Since Darwin* (Norton, 1977). Books on evolution in the series "Readings from *Scientific American*" (Freeman) include *Evolution* (1978), *The Fossil Record and Evolution* (1982), and *Ecology, Evolution, and Population Biology* (1974). On extinctions, there is now a superb volume, *Quaternary Extinctions: A Prehistoric Revolution* (University of Arizona Press, 1984), ed. P. S. Martin and R. G. Kline. Natural selection and evolution is treated in an eminently readable, yet authoritative book by Richard Dawkins, *The Blind Watchmaker* (Longman, 1986).

Local and areal accounts of the Ice Age and its life are published by various institutions and associations. For Beringia see D. M. Hopkins, ed., *The Bering Land Bridge* (Stanford University Press, 1967) and F. H. West, The Archaeology of Beringia (Columbia University Press, 1981). The most famous of all Pleistocene sites is described in C. Stock's *Rancho La Brea: A Record of Pleistocene Life in California* (Los Angeles County Museum of Natural History, many editions). Pleistocene Mammals of Florida, ed. S. D. Webb, (University of Florida Press, 1974) performs a similar function for the entire state of Florida.

Finally, the intermigration between North and South America is treated in *The Great Biotic Interchange,* ed. F. G. Stehli and S. D. Webb (Plenum, 1985).

Index

Argon, 10
Armadillos, 31, 41, 77–79, 80, 129
Armored animals, 77–80
Arredondo (Fla.) site, 69, 70
Artifacts: early man, 133–34
Artwork, 62; *see also* Paintings
Asia, 8, 60
Asiatic lion *(Panthera leo persica),* 59
Asses, 25, 26, 107
Astronomical Theory, 13–14, 86, 144
Atlantic Ocean, 125
Aucilla River (Fla.) site, 70
Auffenberg, Walter, 69
Australia, 8, 53; colonization of, 130, 131–32
Australopithecines, 131
Avifaunas, 29–30
Avocets, 111

Badger *(Taxidea taxus),* 29, 114
Bald eagle, 110
Balkan, 67
Bali, 131, 132
Barbary lion *(Panthera leo leo),* 59
Barriers: ice, 6, 52; water, 31
Bat Cave (Missouri), 95
Bats, 39
Bearded seals, 64
Bearlike dogs, 37
Bears, 76, 80, 82–83; Appalachia, 127; Pleistocene: Great Plains, 98–99
Beautiful armadillo *(Dasypus bellus),* 78
Beaumont, Gérard de, 34–35
Beaver *(Castor canadensis),* 29, 73
Beresovka Expedition, 54
Beresovka River, Siberia, 54
Bering, Vitus, 64
Beringia, 8, 37, 42, 52–54, 114, 129; bears in, 99; elephants' crossing, 43; ground sloths in, 128; horses in, 98; immigrations, 123; life forms, 55–59; lions in, 60; mammoths migrating through, 44; man's arrival in America through, 130; marine fauna in, 63–65; moose in, 123; muskoxen in, 46
Bering land bridge, 22, 35, 53, 124; *see also* Land bridges
Bering Strait, 52, 53
Big Bone Lick (Ky.) site, 58, 119
Binagady deposit, 103
Biostratigraphy, 14–17

Birds: Appalachia, 126; evolutionary history of, 29–30; flightless, 32; Pleistocene: Florida, 80, 83–84; Rancho La Brea, 104; tar pit record, 110–11
Birds of prey, 39, 68, 110, 126
Bison, 38, 41, 45, 80, 87–89, 91, 99, 107, 129; American buffalo: *Bison bison,* 58, 88, 89; Appalachia, 126; *Bison bonasus,* 58; *Bison antiquus,* 88–89, 105; *Bison latifrons,* 88; *Bison priscus,* 58–59; diet, 142; first appearance south of land ice, 50; fossils: Beringia, 54, 58; extinctions, 136, 139–40; hunted by man, 141; Rancholabrean, 59; range of, 61; species, 87
Black bear *(Ursus americanus),* 39, 40, 83, 113, 127
Blancan age, 4, 20, 22–24, 25; animal life, 27–30; climate, 85; duration of, 5; extinctions, 135; intermigration, 52; life forms, 18; species, 47; transition to Irvingtonian, 37, 40
Bobcat *(Lynx rufus),* 29, 83, 87, 101, 107, 114
Bog lemming, 87
Bone-eating dog *Borophagus,* 36, 37, 42, 47
Bones: human, 133
Boreal redback vole, 87
Borneo, 131
Bottlenecks, 141
Bovid family, 45–46
Box turtle *(Terrapene carolina),* 41, 69–70, 84; subspecies *T. carolina putnami, T. carolina carolina,* 69; successor: *T. carolina major,* 69
Brain size: dire wolf, 104–5; and intelligence, 81; large cats, 93, 106; lions, 60, 61–62; llamas, 28
British Isles, 5
British Museum, 121
Broad-fronted moose *(Cervalces latifrons),* 124
Broadwater (Neb.) site, 23
Brodkorb, Pierce, 32, 83–84
Brooks range (Alaska), 52
Brown bear, 61; species *Ursus arctos,* 82
Bruneau (Idaho) site, 43
Brunhes normal chron, 12, 39
Brynjulfson Caves, 105, 125
Bunodont mastodons: genus *Haplomastodon,* 140–41
Bunodonts, 27

Bush dog *Speothos,* 95
Butchering, 139, 140

"Caballine" (true) horses, 26
Caimans, 71, 74, 76
Caliche, 4
California: tar seeps, 102–03
Camelids, 89; *Titanotylopus,* 28
Camelopini *(Camelops),* 29, 89
Camel extinctions, 135
Camels, 41, 87, 129, 133; *Camelops,* 139–40; *Camelops hesternus,* 89, 107; in North America, 28
Canada: glaciations, 5
Canada lynxes, 55
Cape Deceit (Alaska) fauna, 49, 58
Cape Deceit (Alaska) site, 42
Cape hunting dog, 104–05
Capybaras, 31, 47, 73, 76, 129; *Hydrochoerus holmesi,* 72; *Hydrochoerus hydrochoeris,* 71–72; *Hydrochoerus isthmius,* 71
Caracara hawk, 75
Carbon dioxide, 144
Cariamidae, 32
Caribou (reindeer; *Rangifer tarandus*), 56, 57, 58, 129; Appalachia, 126; Beringia, 55
Carnivores, 29, 30, 39, 87; Blancan Age, 37; extinction, 134, 135, 137; North America, 35–37, 129; Pleistocene: Florida, 82–83; range, 61; South America, 32–33
Carpinteria (tar seep), 103
Casper (Wyoming) site, 139–40
Catlike dogs, 37
Cats, 24; marsupial, 30; *see also* Great cats
Cattle, 45
Cave bears, 82; *see also* Florida cave bear
Cave lion *(Panthera leo spelaea),* 59, 60, 62
Caves, 39, 58, 60, 68, 90, 114–15; Appalachian, 125–28; California, 112–14; Irvingtonian fossils, 39–40; Malta, 117; Ozark, 126
Caviomorph rodents, 31, 71
Cenozoic Era, 18
Central America, 31
Central American isthmus, 8
Central Lowland, 85
Chaco (Paraguay) site, 95–96
Chaco Indians, 95
Chamberlin, Thomas C., 9, 12
Chamois, 115

81; extinction, 135, 145; *Glossotherium harlani*, 80, 108–10, 127; hunted by man, 141; *Megalonyx*, 80; migration, 31; *Mylodon listai*, 109; *Nothrotheriops*, 47
Ground squirrels, 42, 101; genus *Spermophilus*, 70
Grubs, 78
Guanaco, 82
Guettard, Jean Etienne, 18
Guidon, N., 133–34
Guilday, John E., 119, 120, 126, 141
Guinea pig, 31, 71
Gulf Coast, 32, 70; "corridor," 70, 72, 80
Gulf of California, 100
Gumbotil, 9
Guthrie, R. Dale, 142, 143

Habitat(s), 30; armadillos, 80; climate and, 141–42; glyptodont, 79–80; moose, 123; muskrat, 47–48; peccary, 127
Hagerman (Idaho) site, 20, 23, 24, 26, 30, 48
Haile (Fla.) site, 69, 70
Half-asses, 98
Half-life, 10, 11
Hallam, Sylvia J., 132
Hallin, Kurt F., 119–20
Hancock, Henry, 102
Hancock Park, 102–104
Hands, 17
Harbor seals, 64
Hares, 29, 33, 54; genus *Lepus*, 45, 47; genus *Sylvilagus*, 47
Harington, C. R. (Dick), 45, 46
Harlan, R., 88, 108, 109
Harrington, Mark, 101
Harvest mouse, 42, 115
Hawks, 110
Hay, Oliver P., 33–34, 36, 90
Hayden, F. V., 27
Hays, J. D., 13–14
Hemiones (half-asses), 25, 26
Hemmer, Helmut, 60, 61–62, 104–105, 106
Hemphillian Age, 20, 23
Herbivores. *See* Plant-eaters
Herds, 92, 93, 97
Hermit sloth, 80–81, 100
Herons, 110
Herz, Otto, 54
Heyerdahl, Thor, 131, 132, 133
Hibbard, Claude, 23, 46, 49
Hide, 17

Holmes, Walter W., 72
Holocene Epoch, 18, 50
Homer, 59
Horns: antelope, 98, 108; bison, 58, 88; muskox, 45–46, 57; organs of display, 16; woolly rhino, 56
Hornsby Springs (Fla.) site, 70
"Horse graveyard" (Hagerman, Idaho), 23, 26
Horses, 33, 41, 46, 80, 81, 101, 129, 133, 139, 141; Appalachia, 126; diet, 142; evolution of, 25–26; extinctions, 135, 143; modern-type, 22; Pleistocene, 107–8; Pleistocene: Great Plains, 97–98; species *Equus conversidens*, 98; species *Equus giganteus*, 98; species *Equus hemionus*, 98; species *Equus occidentalis*, 98, 107; wild, 25, 97–98; zebrine, 26
Horsfall, R. Bruce, 16
Hothouse effect, 144
Howard, Hildegarde, 30, 104
Hunting dog (dhole): *Cuon*, 114–15
Hunting weapons, 136
Hyaenids: Old World, 34
Hyenalike dogs, 37
Hyenas, 36, 103; North America: *Chasmaporthetes ossifragus*, 33–35, 36, 40; Old World, 34–35; Old World: *Euryboas*, 34; Old World: *Hyena borissaki*, 34; Old World: *Lycyaena lunensis*, 34

Iberia, 67
Ibex, 115
Ibises, 110
Ice age, 1, 2, 3, 5–8; beginnings of, in North America, 39; chronology, 9–10; climatic history, of, 9–10, 141–42; climax of, 64; durations of, 10; end of, 135; fauna, 134, 135; return of, 145; theories of, 13–14, 63, 86; transition to Recent epoch, 141; woolly mammoth embodiment of, 57; *see also* Glaciations; Pleistocene Epoch
Ice ages: recurrent, 6
Icefields: North America, 6, 8; weight of, 7
Ice margin, 6–7, 8; Great Plains, 85, 86
Ice sheets, 22, 56, 134; Great Plains, 86; melting of, 145; pushing animals southward, 67; *see also* Glaciations
Ice wedges, 8

Ichetucknee River, 68
Ichetucknee River (Fla.) site, 70
Idaho: Blancan Age fossils, 23, 37
Idaho muskrat, 48
Illinoian Glaciation, 9, 12, 39, 40, 50, 52, 69
Illinoian interstadial, 88
Imbrie, J., 13–14
Immigration: and change in life forms, 47; moose, 123; *see also* Migrations
Immigration routes: man's migration to America, 130–33
Indian elephant (*Elephas*), 42, 43, 44
Indians, 1, 3
Indonesian islands, 131
Inglis (Fla.) site, 35, 40, 80
Insects, 78, 79
Intelligence, 35
Interglacials, 9, 39, 53, 63, 67, 130; Blancan Age, 22–23; climate in, 51; duration of, 14; Florida, 70; later Pleistocene, 50; present, 50, 63, 85, 144; reasons for, 12–14; recurrent, 141
Intermigration, 8, 22; Beringia, 52–54; North/South America, 31; *see also* Migrations
International Rules of Nomenclature, 33–34
Interstadials, 50, 51, 87, 130
Involuntary rafting, 71, 132
Irvingtonian Age, 20, 22, 23, 28, 37, 38–42; Arctic Ocean in, 64; chronology of, 37–38; climate in, 85; extinctions in, 135; micro-mammals in, 49; migrant mammals in, 45–46; species in, 47; transition from Blancan to, 37, 40
Irvington (Calif.) site, 38
Islands: animals on, 136–37; colonization of, 131–33; land mammals on, 115–17
Italy, 67

Jackrabbits, 101
Jaguar, 40, 71, 76, 96, 106, 107, 114; Appalachia, 127; Irvingtonian, 45; Pleistocene: Florida, 82, 83; range, 61
Jaguarundi, 83
Japan: man in, 132–33
Järvi, Antti, 138
Java, 131
Jefferson, Thomas, 58, 95, 120, 121, 127–28

Meade County, Kansas (site), 23–24
Meadow vole (*Microtus pennsylvanicus*), 49, 87
Medicine Hat (Alberta), 42, 134
Mediterranean peninsula, 67
Mediterranean Sea, 67
Megafauna: extinctions, 50, 134–43, 145; of mammoth-steppe, 143
Megalonychid sloths, 128
Megatheriidae, 80–81, 101
Melbourne (Fla.) site, 70
Merriam, John C., 112–13
Mice, 29, 42
Microfossils, 12, 63
Micromammals, 49
Microtinae, 47, 48
Mid-Wisconsinan stadial, 51, 53
Middle Pleistocene Age, 18
Migrations, 22, 37; to America, 129–31; bison, 59; horses, 26; mammoths, 44; natural obstacles to, 67; Old/New World, 8, 33; seasonal, 56
Milankovitch, M., 13
Miller, George, 115
Miller's Cave (Tex.), 78
Miñaca Mesa, Mexico, 35
Mineralization, 68
Miniature horse (*Nannippus phlegon*), 16
Minks, 75
Miocene epoch, 20, 26
Mississippi River, 8
Moa birds, 136
Models, 137; extinctions, 138
Modern horse (*Equus*), 25
Mofres, Duflot de, 102
Moldavian Soviet Republic, 34
Mongolia, 35
Monkeys, 31
Moon rocks, 10
Moonshiner Cave, 3, 15, 20
Moraines, 85
Moose, 61, 123–25, 129; *Alces alces*, 124, 135; migration to America, 130
Mountain deer, 115
Mountain goat *Oreamnos harringtoni*, 101, 114
Mountain lion, 59
Mt. Blanco, Tex., 23, 35, 105
Mount Monadnock (N.H.), 125
Mule deer (*Odocoileus hemionus*), 125
Murray Springs site, 139
Museum of Comparative Zoology, Harvard University, 5
Muskeg, 123

Muskox: Appalachia, 126; *Bootherium bombifrons*, 55, 58, 129; diet, 142; early, 42–46; Irvingtonian, 45–46; modern, 57–58; *Praeovibos priscus*, 46; Rancholabrean, 57–58
Muskrat, 29, 73, 87; Blancan: *Pliopotamys*, 48; and evolution, 20, 47–49; *Ondatra annectens*, 48; *Ondatra idahoensis*, 48; *Ondatra nebracensis*, 48; *Ondatra zibethica*, 47–49; *Pliopotamys meadensis*, 48; *Pliopotamys minor*, 48
Mustelids, 37, 63
Mutation, 21
Mylodonts, 31, 80, 108
Myths, 112, 113–14; about alligators, 74–75; about mammoths, 54, 118; about sabertooths, 106–7

Nansen, Fridtjof, 130
National Museum of Natural History, Washington, D.C., 16
Natural Chimneys (Va.), 126
Natural selection, 21, 44, 131; size in, 117
Neandertal Man, 90, 129, 130, 138
Nebraska, 23, 41
Nebraska State Museum, 45
Nebraskan Glaciation, 9, 10, 12, 39, 44
Neill, Wilfred T., 68, 74–75
Nevada, 24
New Guinea, 131
New Paris No. 4 sinkhole, 126–27
Nine-banded armadillo (*Dasypus novemcinctus*), 77–78
Nordenskiöld, Erland, 109
North America: glaciations and interglacials, 5, 8, 50–51; ground sloths, 128; horse evolution in, 25–26; hyenas in, 33–35; mammoths in, 43–46; original homeland of camel, 28; Pleistocene fauna, 129; richness of flora and fauna in, 67
Northern Channel Islands, 115–17
Northern Yukon Research Programme (University of Toronto), 134
North Pole, 14
Northwest Territories, 52
Nowak, Ron, 114

Oakley, Kenneth P., 56
Ocelot (*Felis pardalis*), 70, 83

O'Connell, James F., 131, 132
Ogallala Formation, 4
Ohio Mastodon (*Mastodon americanus; Mammut americanum*), 119
Oil, 103
Okapi, 94–95
Oklahoma, 41
Old Crow River basin (Yukon), 134
Olduvai Gorge, 40, 60
Olduvai subchron, 40
Old World: antelopes, 45; glaciations, 5; mammals, 44, 45; moose ancestral forms, 124; wild horses, 25
Old World species: in Beringia, 55–56
Onagers, 98
Opossum (*Didelphis virginiana*), 29, 30, 31
Orr, Phil C., 115
Otavalo Man, 133
Overkill theory, 136–41
Owen, Richard, 108
Owls, 39, 84, 101, 111, 126
Ozark caves, 126

Pacific Ocean, 133
Paintings, 134; Paleolithic man, 58; Stone Age, 56
Pair-bonding, 93
Paleocene epoch, 25
Paleoclimatologists, 63
Paleoindians, 1, 133–34, 136; artifacts, 121; colonization of America, 130–31; and overkill theory, 136, 138, 139–41, 143
Paleolithic man, 58, 132
Paleomagnetic subchron, 39
Paleontologists, 2, 16, 18
Paleontology, 16; synthesis in, 35
Paleozoic Era, 125
Palo Duro Canyon, Tex., 3–5, 6, 14
Palo Duro Creek, 4
Pampatheres, 78–79; North American: *Holmesina septentrionalis*, 79
Pandas, 37
Panther, 59
Passenger pigeon, 111, 136
Past: reconstruction of, 1–2, 41–42
Patagonia, 109
Pearlette Ashes, 23–24
Peccaries, 24, 33, 80, 87, 95–97, 114; Appalachia, 126; extinction, 135; genus *Catagonus*, 95; *Platygonus*, 126, 127; genus *Tayassu*,